Construction Guide
for
Soils and Foundations

Construction Guide for Soils and Foundations

SECOND EDITION

Richard G. Ahlvin, P.E.

Vernon Allen Smoots, P.E.

A Wiley-Interscience Publication

JOHN WILEY & SONS

New York · Chichester · Brisbane · Toronto · Singapore

rec'd 1/9/89

Library of Congress Cataloging in Publication Data:

Ahlvin, Richard G. (Richard Glen), 1919–
Construction guide for soils and foundations.—2nd ed. / Richard G. Ahlvin, Vernon Allen Smoots.
p. cm.—(Wiley series of practical construction guides)
First ed. by Gordon A. Fletcher and Vernon A. Smoots.
"A Wiley-Interscience publication."
Bibliography: p.
Includes index.
ISBN 0-471-80486-X
1. Foundations—Contracts and specifications. I. Smoots, Vernon Allen. II. Fletcher, Gordon A., 1900– Construction guide for soils and foundations. III. Title. IV. Series.

TA775.A433 1988 87-34639 624.1'5—dc19
ISBN 0-471-80486-X

Printed in the United States of America

10 9 8 7 6 5 4 3 2 1

Series Preface

The Wiley Series of Practical Construction Guides provides the working constructor with up-to-date information that can help to increase the job profit margin. These guidebooks, which are scaled mainly for practice, but include the necessary theory and design, should aid a construction contractor in approaching work problems with more knowledgeable confidence. The guides should be useful also to engineers, architects, planners, specification writers, inspectors, project managers, superintendents, materials and equipment manufacturers and, the source of all these callings, instructors and their students.

Construction in the United States alone will reach $270 billion a year in the late 1980s. In all nations, the business of building will continue to grow at a phenomenal rate, because the population proliferation demands new living, working, and recreational facilities. This construction will have to be more substantial, thus demanding a more professional performance from the contractor. Before science and technology had seriously affected the ideas, job plans, financing, and erection of structures, most contractors developed their know-how by field trial-and-error. Wheels, small and large, were constantly being reinvented in all sectors, because there was no interchange of knowledge. The current complexity of construction, even in more rural areas, has revealed a clear need for more proficient, professional methods and tools in both practice and learning.

Because construction is highly competitive, some practical technology is necessarily proprietary. But most practical day-to-day problems are common to the whole construction industry. These are the subjects for the Wiley Practical Construction Guides.

M. D. MORRIS, P.E.

Preface

This book is one of the Wiley Series of Practical Construction Guides. Its aim is to provide practical and useful information in an accessible form. Currently available books on soils and foundations have been written by professors as college textbooks. Such textbooks serve to train engineers and other professionals, but do not supply the information needed by those who will physically construct the foundations.

Part I of the book presents an easy-to-read description of soil and soil behavior. It is intended for busy contractors, not expert in soil mechanics, who must deal with soils during construction. It is also intended for use in acquainting inspectors, learning soil technicians, and soil laboratory personnel with the nature and behavior of soils.

Part II of the book discusses aspects of soil, soil behavior, and foundations commonly encountered in construction. It emphasizes that soil is as much a part of the overall structure as is the concrete, steel, and wood superstructure.

Many of the ideas in the book are based on experience, including the descriptions of job problems and their solutions, and of things to watch for on the job.

A book of this nature seems especially pertinent in the current changing and increasingly complex construction environment. Limited space in populous areas, difficult transportation construction in remote areas, and efforts to protect and maintain the environment, along with increasing ability to deal with soil difficulties, have led to increased use of less desirable earthwork and foundation sites. As a result, construction is more complex, and contractors must exercise greater awareness and ingenuity in their operations.

Thanks are due those who contributed to the preparation of this book. The Series Editor, M. D. Morris, encouraged its writing. Mr. G. B. Mitchell, P.E., Chief of the Engineering Group, Geotechnical Laboratory of the U.S.

Army Engineer, Waterways Experiment Station, reviewed the revised manuscript and made valuable suggestions.

Recognition is due Gordon A. Fletcher (deceased), co-author of the prior book, *Construction Guide for Soils and Foundations* by Fletcher and Smoots, from which this book is a substantial revision. Much of the material in Part II of this work is taken directly from that earlier text, and the many contributors to items in that earlier book are again due thanks without specific recognition here.

RICHARD G. AHLVIN
VERNON ALLEN SMOOTS

Vicksburg, Mississippi
Los Angeles, California
April 1988

About the Authors

RICHARD G. AHLVIN has enjoyed a career predominantly in research in relation to behavior of soils. His schooling ranged from a structural civil engineering degree, through much of a degree in mechanical engineering, an advanced degree in highway engineering—all from Purdue University—to aviation engineering cadet training in the military, an advanced soil mechanics course from MIT, and to the teaching of elementary and advanced surveying, testing materials, and engineering mechanics at Purdue and of soils for transportation facilities at the Mississippi State University Extension at Vicksburg.

From early work in the field and laboratory supporting studies of vehicle trafficability on soft soils through midcareer work in research of soils for subgrades for heavy duty airfields, drainage installations, and unsurfaced landing fields, Mr. Ahlvin's career led to Assistant Chief of the Corps-of-Engineers Geotechnical Laboratory with responsibilities for all aspects of civil and military soil-related problems and doctrine.

In a second career, following formal retirement from the Corps-of-Engineers, Mr. Ahlvin has been involved with soil problems and heavy airfield pavement performance in East Malaysia, Kuwait, Ireland, South America, and various U.S. locations.

VERNON ALLEN SMOOTS (AL) is a retired partner of Dames & Moore, geotechnical consulting engineers headquartered in Los Angeles. Associated with Dames & Moore from 1946 to 1985, Mr. Smoots managed the New York office before becoming Managing Consultant of the Los Angeles office in 1954. As a consultant in soils and foundations with engineering registrations in 11 states, his projects include flood control, water transmission, dewatering of excavations, oil refineries, power plants, recreational facilities, and coastal and offshore structures in the United States and overseas. He holds a B.S. from the University of Kansas (1944) and served two years as a lieutenant in the U.S. Navy Seabees. He is a Life Member of ASCE and a contributing author to ASCE's Construction and Transportation Journals.

Contents

PART II APPLICATIONS

List of Illustrations

Construction Guide
for
Soils and Foundations

I
Concepts

1

Introduction

1.1 Purpose of the Book

Earth materials, primarily soils but also rock, probably are the most common and least well understood of the materials with which a contractor must deal in construction. Accordingly, the purpose of this book is to convey concepts of the behavior to be expected of earth materials when dealt with in the field by contractors, inspectors, and others.

There is no intent to compete with the formal geotechnical text references of soil and rock mechanics and foundation engineering. These textbooks remain the basis of university instruction and source material for sophisticated design analyses. Most theory has been left to such sources, and emphasis herein is placed on practical applications. A special effort has been made to simplify concepts and promote understanding with a minimum of reader effort.

It is expected—and strongly recommended—that for extensive, involved, costly, or difficult projects a contractor take advantage of established geotechnical expertise to assist with any soil, rock, or foundation problems encountered. On the other hand, it will commonly be necessary for contractors and others to deal with their own problems on smaller, less well budgeted projects, or to recognize when it becomes necessary to call in special expert assistance. This book will help with such needs.

Recognizing that geotechnical matters will commonly not be a primary concern of readers and may be an unfamiliar area, the first part of the book will deal with concepts of soil behavior and will assume interest and intelligence but no significant knowledge of the subject. The second part of the book will deal more specifically with matters of actual practice.

1.2 Earth Materials

Concern here will be with the mechanical behavior of surface and near-surface soils and rock. Deeper materials and, most commonly, rock are the realm of the geologist, who can frequently provide valuable information also on behavior of the shallower materials. The agricultural behavior of soils is the realm of the soil scientist, from whom we also commonly borrow. Both of these sciences enjoy much older and broader backgrounds than soil mechanics does, which, while having some older roots, has essentially developed as a science within this century. Rock mechanics developed even later, virtually within the last half of this century. The term geotechnical is now used to refer to soil mechanics, rock mechanics, and all aspects of foundation behavior.

1.3 Soil Formation

Basically soils derive from rock ground down by water, wind, and wear. Wetting and drying, heating and cooling, freezing and thawing, and pressure of root growth all tend to fracture and scale the base rock. Flowing water, wind, rock slides, glaciers, even animal activity all wear down the rock and contribute to the soil-forming process.

Loosened or broken source materials are repeatedly transported, mixed, deposited, and sorted by water, wind, gravity, glaciers, volcanic activity, etc., and in the process they are further abraded and ground to smaller particle sizes. Mixing results from one source material being moved to join another or from two or more materials being intermingled during the transport process. Successive movements multiply the process. Sorting is predominantly related to the force of the medium causing movement. White water flows can move boulders, fast water can move sand and gravel, slow water moves fines, and quiet water permits transported fines to settle. Similarly, wind moves fines, as dust, over distances for deposition in quiet air, and it can roll and bounce sand sizes to accumulations in dunes and other deposits. Slides and rockfalls cause some segregation.

Additives are contributed in the soil-forming process primarily by plants and animals. In more recent times humans have also contributed to a degree, through their agricultural practices, manufacturing operations, and waste disposal.

Chemical modifications of soil materials occur through leaching of even poorly soluble elements and compounds; through combining with porewater-borne chemicals; through combining of materials brought together by wind, water, gravity, glacier or volcanic activity; and through additions and depletions resulting from plant, animal, and human activities. Physical and physiochemical changes result from downward flow, leaching, and re-deposition action of rainfall. In arid or periodically arid areas, surface evaporation and upward (recharging) flow of moisture results in accumulations of precipitated chemicals in soils.

1.4 Consistency in Soil Development Processes

There can be wide variety in each of the elements of soil formation described, from the type of source rock through the many processes of breakdown, mixing, transport, additions, and chemical change and the repeated recurrence of each. Despite this potential for infinite variety there are fairly strong consistencies in the soil development processes, and by observing these much can be learned of the nature of soil at a particular site.

Searching studies have been made of these processes in relation to physiographic development, glaciation, and young and mature river patterns. Methods for air photo interpretation recognize these and other processes and study agricultural practices and natural vegetation in relation to soils and the impacting environment. While these sources of site soil data can be interesting and in some circumstances valuable, they are generally beyond the scope of this presentation and will commonly not be an efficient source of soil information for individual construction sites. For any site, however, obvious and readily available information should be considered.

Steep upland streams and near mountain rock sources do not tend to development of fine soil deposits and mature, greatly modified soils. Only fine deposits exist in the lower reaches of mature rivers; sands remain in bars while flowing water removes the silt and clay sizes, which are deposited in quiet water in deltas, lakes, and cut-off oxbows. Freshwater silts are flocculated by salt when carried into salt or brackish water, and the heavier flocs settle out into high-void accumulations, which are usually highly compressible.

The greatest modifications to soil formation occur in high-rainfall tropic areas as a result of plant and animal contributions, leaching and chemical changes, and physical and physiochemical changes. In arctic areas formation processes are largely limited to mechanical breakdown, transport, mixing, and sorting.

No comprehensive descriptions of these processes can be given here, but the reader is encouraged to contemplate the natural physical processes in relation to a particular site soil assessment. Much can be learned or deduced by observing such elements as the general soil type, uniformity of site soils, vertical and lateral variations or changes in soil type, wet and dry soil conditions, possible shallow groundwater, and shallow rock.*

* The reader should be warned that world climatic changes, continental drift, and shifting of the earth's axis over vast time periods can produce geologic surprises. Such past occurrences have resulted in soils developed under tropic conditions now existing in temperate zones, temperate zone soils now existing in arctic zones, etc.

2

Regarding Soil

2.1 Nature of Soil

Despite consistencies of developmental patterns, there is wide variation in soil source materials and forming processes, and therefore soils do vary through the entire range of gradations, plasticities, etc., and deposits can yield uncommon or unexpected behavior characteristics. In dealing with soils one must always be wary of and prepared to cope with any unusual behavior encountered; however, much of the behavior to be expected of a soil can commonly be deduced from its general nature or classification as sand, silt, clay, or some combination or variant.

Clay is composed of very fine particles (less than about 0.002 mm in largest dimension), retaining few if any of the characteristics of the rock from which it originated. Since breaking any particle adds two new surfaces with increase in surface area, it will be readily appreciated that the total surface area of the very fine particles making up a clay soil is enormous. The effects of water adsorbed (a dispersed electron bonding, *1**) on particle surfaces and of capillary action of water absorbed in the fine pores between particles have great influence on the behavior of clays. The finer clay particles (less than about 0.0002 mm) are almost entirely of "platey" crystalline structure having a surprisingly small variety of crystalline and chemical compositional forms, *2*. Most clay minerals fall into only three crystal structures, with some chemical variants, but there is wide variation in surface and crystal lattice moisture effects, which results in great behavioral differences. Clays are characterized by plastic (putty-like) behavior of greater or lesser degree depending on consistency and composition.

* Italicized numbers refer to items in the list of References.

Sand is the material of beaches, dunes, river bars, etc. Individual grains can be easily discerned (particles passing a no. 4 sieve and retained on a no. 200 by the USCS (Unified Soil Classification System) *3, 4*). Typically sands with few or no fines (where "fines" are material passing a no. 200 sieve) are nonplastic and little affected, as to behavior, by particle surface area phenomena. Thus their behavior is predominantly mechanical, except with regard to pore water capillarity in the moist-to-damp condition and pore water pressure (in some dramatic circumstances the quicksand condition) in the saturated state. Sands are, of course, not limited to the characteristic depositions indicated above but are commonly encountered in surface and subsurface strata, particularly in coastal plain areas, in mountain outwash areas, in sandstone country, etc. Initially, the behavior of sands will be discussed in relation to the low-fines, nonplastic materials described, but it must be noted that significant to substantial (10–50%) fines can greatly influence behavior. Hopefully this will become clear with subsequent discussion.

Silt is composed of particles too fine to be considered sand but which are not greatly modified by chemical and physiochemical processes and thus not reduced to clay minerals. Generally, silt particles are the coarser of the fine soils, and some classification systems attempt to separate silt from clay on the basis of grain size alone (0.002 mm is used, but some separate at other sizes).

The USCS *3, 4*, separates coarse soil from "fines" on the basis of particle size (fines pass a no. 200 mesh sieve; coarse soils do not), but divides silt from clay (fines) on the basis of plasticity characteristics. Silts show no to some plasticity, while clays show medium to high plasticity.

The characteristic behavior of sands, silts, and clays will be described, but first the effects of moisture and plasticity, and soil density will be examined.

2.2 *Moisture and Plasticity*

The consistency of soils and therefore their behavior is greatly affected by the moisture condition. Fine soils tend to softness and lower strength at higher moisture contents and to firmness and higher strength at lower moisture contents. To understand this concept, consider the entire moisture range, from soil particles dispersed in water to a dry, solid mass or mass of accumulated soil grains. At the wet end of this moisture range, soil exists as a slurry. It is a viscous liquid in which the particles are dispersed or "floating" in the water. Particle interactions are by "bumping" contacts (or complex particle-surrounding force-field interactions not well defined) which increase as moisture is reduced. Viscosity increases and at some point the consistency passes from that of a liquid to that of a plastic mass. The moisture content at this point of change is defined as the "liquid limit" (LL), and simple standard tests, *5*, are employed to establish it.

Soil in a plastic mass condition is putty-like. It is readily deformed and reshaped. The moisture present lubricates the soil particles, permitting them to slide over one another with ease. At the wetter end of the plastic range, nearer the liquid limit, soils are soft and easily extruded, much like toothpaste or caulking compound. At the drier end soils are tougher and more like a child's modeling clay or caramel candy. The moisture content at the point of change from a plastic mass to a semisolid is defined as the "plastic limit" (PL), and standard tests, *5*, are employed to establish the PL.

The width of the moisture range from wetter LL to drier PL in the plastic mass condition is the "plasticity index" (PI), and it indicates the degree to which a soil exhibits plasticity in consistency and behavior. Thus, LL − PL = PI. Low PI indicates low plasticity while high PI implies highly plastic behavior.

Further drying, below the PL, brings the soil into the semisolid mass condition. Particles are in contact and will not easily move with respect to one another. However, water remains in the crystal lattice of (expanded) clay mineral particles; it also remains bound or adsorbed on the surface of very fine particles and absorbed in the capillary pore spaces between particles in the coarser of the fine-grained soils. Thus soils in the semisolid mass condition are subject to shrinkage with drying. The dry-side limit to this condition is defined as the "shrinkage limit" (SL), and is also established in terms of moisture content by a standard test, *5*.

At a moisture dryer than the SL, soils are in a solid mass condition. In this condition soils appear to be dry and retain little moisture. They are subject to no further shrinkage, and excepting "clean" sands—sands with no appreciable fines—the soil particles in a solid mass condition are cemented or bound together, the strength of the bonding depending on the degree of plasticity of the soil.

Figure 2.1 shows the general pattern of soil consistency for the entire range of moisture, from dry to liquid, and for the full range of common soil types, soil grain sizes, and plasticity. Again, it must be understood that the representation is broadly and generally applicable, but cannot adequately include many exceptional and special-case soil materials.

2.3 Natural State Soils: Consistency, Density, and Voids

Soils can be found in nature existing in any of the consistency conditions described in Section 2.2. Freshly settling silts and clays are found in new delta deposits, around drying lakes, etc. Arid region deposits can be almost completely dry. However, upland soils, both surface and subsurface, are commonly found in only limited ranges of the moisture/consistency array. These soils experience grain-to-grain contact and have been subjected to some consolidation either by past geological processes or merely by wetting

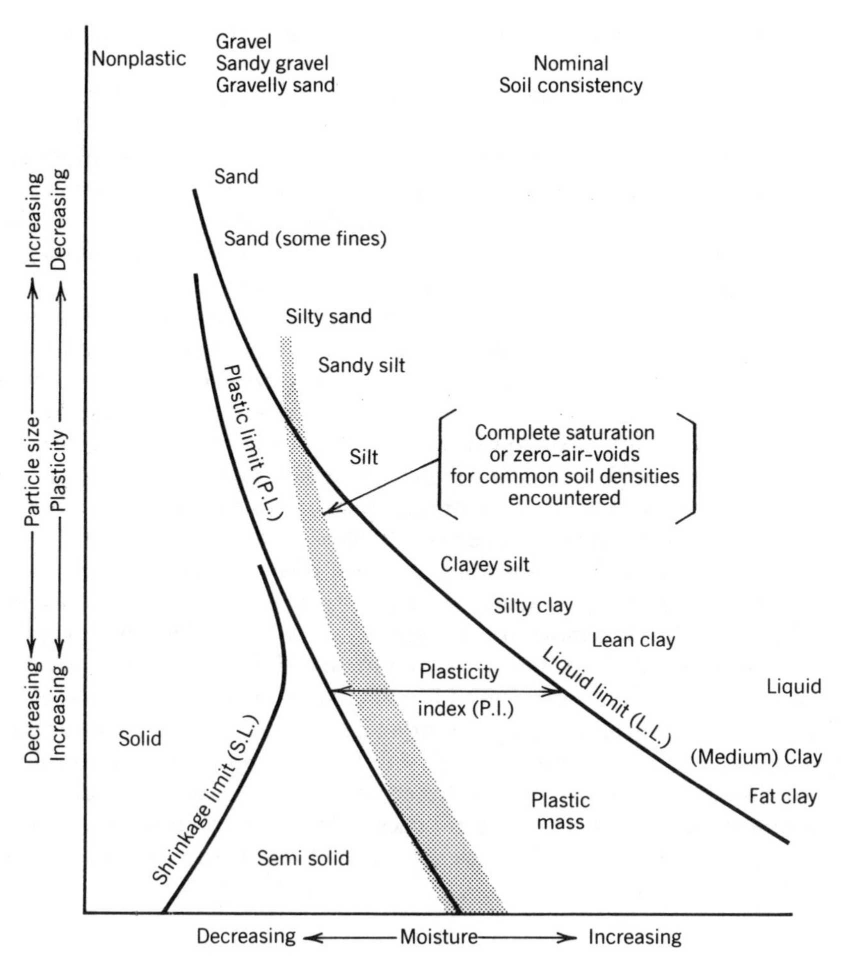

Fig. 2.1 Nominal soil consistency.

and drying, heating and cooling, freezing and thawing, overburden weight, surface pressures, etc.

Thus any particular type of soil will commonly be found to exist within a limited range of density. There are exceptional soils to be discussed later, which have uncommonly large pore space within the solid grain-to-grain structure and others which have a capacity to swell greatly with wetting and to shrink on drying. But in general a soil has only limited pore space for water to fill and will swell and shrink only a little on wetting and drying. So, excepting the unusual soils, any soil encountered will exist and tend to remain at a fixed density with a fixed amount of pore space between soil grains, and the range from lowest to highest density to be encountered will generally be rather limited.

To further clarify this concept and to provide a basis for a quick consideration of some attributes of soil, consider soil as a volume filled with soil particles in contact, with voids or pore space between grains. Water can fill the voids and be withdrawn from them by drainage, evaporation, etc. But soil moisture will be limited to the voids-full or zero-air-voids condition of a soil, and higher moisture contents can only obtain by physical reworking of the soil to a lower density and higher voids condition.

Commonly the finer soils—clays and fine silts—exist at densities ranging from about 85–105 lb/ft^3, while coarser sands and gravelly sands range in density from about 105–125 lb/ft^3. Intermediate sizes and mixtures will be found between these ranges. The stated densities are dry densities, or the unit weights of volumes of soil from which all water has been removed.

For these conditions, ranging from less dense finer soils to more compact coarser soils, and considering a common mass-density of the soil solids, the solids will occupy from 50% to perhaps 75% of the volume. Thus in a cubic foot of soil the pores between soil grains will provide space for only about 30 lbs of water at lower densities and for about 15 lbs of water at higher densities. Considering moisture content, w, the weight of water divided by the weight of solids for a given volume of soil, the highest, or saturation, moisture content commonly found in low-density clays or fine silts is about 35%, while for dense and coarse sands and gravelly sands, the saturation moisture is at limits as low as 12%. Figure 2.1 also shows the common nominal limits of moisture for saturated or zero-air-voids soil conditions in relation to soil consistency characteristics. Here again it must be cautioned that specific limits such as these are presented for guidance, and exceptions will often be found.

3

Soil Behavior

3.1 Characteristic Behavior of Sand, Silt, and Clay

First consider sand as it is commonly found on beaches, in dunes, in river bars, and wherever flowing water may deposit sand-size grains while removing finer silt and clay-size particles. This type of sand deposit is commonly encountered in excavations and foundation explorations in thick and thin layers formed as ancient beach ridges, in river valley deposits, and in mountain outwash areas.

Sands with no fines are nonplastic, exhibiting no plasticity characteristics, and they are cohesionless, with no tendency for grains to cohere or attach to one another. Some exception to this is provided by capillary action of water in the pore spaces between grains, as will be seen later, but initial considerations will be of nominally dry (or completely saturated) sands. With no capacity for grains to attach or hold to one another, sands exhibit instabilities and cannot stand in side cuts or embankments at angles above about 35° to 40° to the horizontal. They are also subject to displacements underfoot, under shallow foundations, or under the tires or tracks of vehicles.

Resisting these displacements is a matter of resisting internal shearing or, as has become accepted terminology, internal friction. Consider for example two pieces of sandpaper placed with the sanded faces in contact. Sliding one piece with respect to the other meets with little resistance if the pieces are not pressed together. If, however, a weight is placed on the face-to-face pieces of sandpaper, resistance to sliding one over the other is substantially increased, and the greater the weight, the greater the resistance to sliding. Shearing or internal friction is a function of the force normal to the plane of shearing or friction. In relation to a volume of sand it is a function of the confinement of the sand volume or of the pressure pressing sand grains together. Friction between two surfaces involves a friction coefficient which depends on the character of the surfaces in contact. Likewise the internal friction within sands is expressed as an "angle of internal friction" whose tangent is equivalent to a friction coefficient. Typically the "friction angles" for sands can range from below 30° for rounded, fine, one-size sand particles to above 40° for angular, mixed-size, coarse sand particles. Thus dry sands with few fines are characteristically of poor stability when loose and unconfined but are brought to good stability and shear-resistant strength by confinement. This stability is dependent primarily on the amount or degree of confinement and secondarily on the irregularity of potential internal shear surfaces as controlled by particle angularity, particle shape, and the mixture of diverse particle sizes.

Damp and wet sands, structured lower density sands, and sands mixed with silts and clays will be further discussed after the characteristic behavior of silts and clays is examined.

Consider clay next. Some clays at suitable moisture conditions are puttylike or similar to a child's modeling clay, but both the basic characteristics of the clay materials and the moisture conditions, ranging from wet to dry, can vary widely. Thus clays vary in behavior and character. Clays are common in surface and subsurface deposits, in ancient and recent deposits, in lacustrine and mature river environments, in shale and other residual soil areas, and where soil-forming processes have long or intensively been active. In some geologically young outwash areas, arid climate areas, or arctic areas, clays can be virtually nonexistent.

Clays are plastic; some have limited plasticity while others are highly plastic. They are cohesive; grains have a strong tendency to cohere or attach to one another. Because their soil particles are bonded together, clays have inherent stability or resistance to shearing determined by their plastic nature and by their consistency with respect to moisture condition.

Clays, except when near or above their liquid limit moisture content, will stand vertically in sidecuts or excavations, but only to a certain height, depending on their shear strength. At this height the weight of the adjacent soil will cause collapse of the cut-face. This can happen without warning and is the cause of many construction catastrophes. It is thus very important for anyone dealing with vertical sidecuts, excavations, and particularly ditches to heights above 4–5 ft to be quite sure of safe heights before proceeding without bracing and shoring.

The strengths or stabilities of clays—their resistance to internal shearing—are largely to entirely dependent on their plasticity or cohesive internal bonding. Unlike sands, highly plastic clays gain their shear strength entirely from this cohesive bonding, without regard to the presence or absence of any confinement of the soil or normal pressure on potential shear surfaces. Leaner, less plastic clays gain their shear strength in part from cohesive bonding and in part from internal friction dependent on confinement. It follows that the "fatter" or more cohesive the clay, the more it depends on cohesive bonding for strength, while the "leaner" or less cohesive the clay, the more its strength depends on internal friction and confining stresses. Friction angles for lean clays range from a few degrees to perhaps 20° or above. The leaner the clay, the higher the friction angle. Fat clays, as already mentioned, have friction angles of zero to only a few degrees.

On the plasticity scale, as clays range from wet to dry (through LL, PL, and SL), the range in moisture content is broad. Generally, in natural deposits clays do not have sufficient voids to accommodate moisture contents much above the PL. Only very recent deposits in forming deltas, embayments, and lakes, or sensitive clays (discussed later), or some very lean clays with low LLs will normally have voids sufficient to accommodate saturation moisture near or above the LL. Thus, although clays are not as strong as less plastic soils, they do not commonly exhibit wide variations in strength, from moist to saturated, in-situ.

Silts can, in many ways, be considered to fall between sand and clays, yet a too simplistic acceptance of this idea can lead to problems. Well-understood and properly controlled silts can be considered not as strong as confined sands but stronger than soft clays. Some soil classification systems attempt to separate sands, silts, and clays on the basis of particle size alone. Yet loose sands are less stable than silts, and plasticity of the very fine particle soils is far more significant than grain size in determining behavior.

Silts are widely considered to be risky and treacherous, not in a "quicksand" sense, but because their behavior is quite often less than expected or intended and they can exhibit dramatic strength loss in certain circumstances. The grain size and pore space of silts are such that they readily respond to moisture change on wetting and drying. Yet they strongly retain capillary water and can restrict flow and be difficult to drain. Their reputation for difficult handling and tricky behavior results from their low plasticity and common in-situ density ranges. Their commonly existing natural densities provide pore space which will accommodate moisture ranging from dry to, sometimes, above their liquid limit. With pore spaces nearly to entirely full of water, yet with soil grains in contact, silts can exhibit a measure of strength or resistance to loading. But with repeated loading, kneading, or manipulation, the grain contacts are reduced through pore-pressure buildup (or other not yet thoroughly explained phenomena), with a resulting dramatic loss in strength or resistance to internal shearing. This is similar in some ways to "sensitivity" of high void-ratio clays, and is sometimes confused with sensitivity behavior. However, it is not the same phenomenon, since the in-situ stability can reestablish in silts in a relatively short time while clay sensitivity is the result of geologic processes.

Another reason silts have a reputation for tricky behavior relates to the difficulty in making use of them at their best strength. Silts—and even more so clays—have irregular particle shapes which, when randomly distributed within their soil mass and compacted under moisture conditions somewhat below saturation (on the dry side of optimum, see Section 12.9 and Fig. 22.2), apparently have their randomly oriented grains jammed together. This results in a relatively strong shear-resistant condition. To understand this concept, consider tamping together randomly distributed masses of nails or toothpicks. The same soils when compacted or when manipulated in-place by shearing or kneading at moisture contents nearer to saturation (wet side of optimum) apparently permit grains to assume less random orientations with resultant reductions in shearing resistance or strength.

The treacherous aspect of silts is due to the quite narrow range of moisture at which they can be compacted to their highest strength. The range of moisture is only a few percent and thus much narrower than commonly exercised in construction control of moisture. Drier compaction can appear to result in good strength, but the strength is lost on wetting. Wetter compaction results in greater particle orientation and lower wet strengths.

3.2 Mixed Soil Types

Sands, silts, and clays exist in nature as individual soils, but mixtures of the three, or any two of the three, are much more the norm. The common agricultural term for mixtures that nicely support plant growth is loam. Although engineering aspects of soil behavior must recognize references to loam, sandy loam, silt loam, clay loam, etc., for the information they may convey, no engineering soil classification finds loams meaningfully definable. The collective characteristics of such mixtures represent too broad a range of behavioral attributes for engineering purposes.

The prior discussion of sands was concerned with "clean" sands—those without fines. Errors in judgment of expected behavior of sandy soils are not uncommon when the sands are "dirty"—mixed with fines. The soils can be obviously predominantly sand, yet mixed with fines such soils can have their behavior primarily determined by their fine fraction. Silty sands, silty-clayey sands, and clayey sands can behave much more like silts and sometimes more like clays than like the clean sands described earlier. It is common to require no more than 15% fines (particles passing a 200 mesh sieve) in sandy soils intended to behave as sands, with the additional requirement that the fines be of low (less than 6 PI) to no plasticity. As little as 10% or even less can have a great influence on behavior if the fines are of higher plasticity.

Regarding mixtures of silt and clay, which commonly have appreciable percentages of sand, such soils have their behavior predominantly determined by their plasticity. This is why engineering soil classification systems attempt to separate silts from clays on the basis of plasticity (see ASTM D-2487, *4*).

As mentioned earlier, there are soil classification systems—especially agricultural and geological systems—which attempt the silt and clay separation on the basis of particle size alone. It is also common practice in the field to identify soil type on the basis of "visual-manual procedures" (see ASTM D-2488, *7*), without specific plasticity or grain-size determinations. Thus field logs of soils investigations commonly include references to sandy silts, sandy clays, sandy-silty clays, clay silts, silty clays, etc. Such logs have primary value in connection with construction operations involving the soils logged, but some care must be exercised in judging just how well the soils logged have been identified.

It would be well to note, at this point, that the term "sand," for example, refers to soils in which sands predominate, while at the same time sand is separated from other sizes in classification systems (for example in the USCS ASTM D-2487, *4*) as the soil that contains grain sizes passing the no. 4 and retained on the 200 mesh sieve. A soil classified as sand (USCS) can have as little as 26% of grains in the size range considered to actually be sand. Up to 49% can be fines and up to one half the remainder, or say 25%, can be gravel sizes. To be sure, an attempt is made to reduce confusion with qualifying adjectives, (e.g., silty-clayey-gravelly sand or clayey sand with gravel), but basically these are all called sands.

3.3 Gravel and Other Coarse Aggregate

The USCS, *4*, divides gravel from sand on the no. 4 sieve. Other soil classifications separate sand and gravel sizes but not necessarily at exactly the no. 4 size. Gravels have the same terminology problem that sands do. For size separation the USCS, *4*, divides gravel from cobbles on the 3-in. opening sieve and cobbles from boulders at the 12-in. size.

Gravels exist in sorted deposits in upland areas near the source rock where rapidly flowing water has provided the "grinding," movement, and sorting. Commonly gravels are found in river terrace deposits but can be encountered in buried layers, as a result of sheetflow deposition. In glaciated areas and meltwater outflow valleys, deposits of sorted gravels are to be found well away from mountain rock sources.

Mixed soils including portions of gravel and coarser aggregates exist in areas where the source rock is being broken down, in outwash areas where braided stream and sheetflow action result in movement and mixing, and in glacial tills where grinding, mixing, and deposition result in mixtures of all size soil particles. The term "boulder-clay" is employed to designate glacial mixtures of clay-size to boulder-size particles.

The worn condition of most gravel particles results in the well-rounded rock pieces commonly expected of gravels, but less worn, angular "broken rock" deposits can be encountered. Man-made crushed rock can have sizes from gravel to sand to fines and can consist entirely of crushed "ledge" rock or can consist of crushed gravel, cobbles, or boulders. In the latter case not all particle faces will be "split" or "crushed" faces, and "angularity" or particle interlock can vary.

Clean gravels are, of course, cohesionless, like sands, and the behavior of dirty gravels is much affected by the quantity and plasticity of fines. Multiple points of contact between grains, with confinement pressing these contact points together, provide good stability. The presence of fines and especially of plastic fines with low total-mass moisture but relatively high moisture in the fines can lubricate coarse grain contacts and reduce stability. With sufficient fines the contacts between coarse grains are reduced. When fines (−200 materials) and sand exceed about 50% of a gravelly soil, the soil tends to behave as though the gravel sizes were not present.

3.4 Some Special Behavior Concepts

The effect of capillary fringes between grains in moist sands was mentioned earlier. These fringes provide a bonding or cohesive effect commonly termed "apparent cohesion," since it represents behavior in clean or fines-free sands much like the true cohesion of lean clays. This binding provides the measure of stability to be observed at the water edge on beaches. Apparent cohesion is lost in either dry or saturated sands. In gravels or very coarse sands the capillary effects are small or insignificant, but for medium and especially fine sands it can be quite significant. At moistures intermediate between dry and saturated the capillary fringes lead to a condition called "bulking" in which the bonding between grains rather strongly retains a higher voids or "bulked" condition. The bulking condition commonly leads to support problems in too casually treated construction applications, because the apparent cohesion resists compaction and voids reduction, while later saturation eliminates the cohesive bonding. The result is unwanted settlements or voids beneath pavements, floors, footings, etc.

Quicksand is broadly known but commonly misunderstood. It is not, of course, the sand but a condition of the sand. Sands loosened by upflowing water or laid down in a very loose, high-voids condition, yet a condition in which there is grain-to-grain contact and some stability, will become "quick" when completely saturated and disturbed by shock or sudden loading. The sudden loading or shock applies pressure to the pore water, causing the water pressure to separate the sand grain contacts. The result, for a brief period, is a liquid filled with dispersed sand grains, which provides little or no support for applied loadings.

Worth mention here is a sometimes valuable device for examining expected behavior of gravelly and sandy soils containing appreciable quantities of plastic fines. This is only possible when the plasticity characteristics (LL, PL, and PI) of the fines have been determined or can be reasonably closely judged. It must also be possible to decide the proportions of coarse and fine material present in the soil. Common practice provides the moisture content of the total soil. The proposed device determines the total quantity of water in the soil from the reported percent moisture, allowing a reasonable (2–3%) amount for absorption of water in the surface pores of the coarse particles, and determines the effective moisture content of the fine fraction, assuming all but absorbed water to be in the fine fraction. Comparison of this moisture content to the LL and PL gives an indication of the behavior to be expected from the fines filling space between coarse particles.

4

Unusual Soils

4.1 Swelling Soils

All plastic soils are subject to some swelling on wetting and drying, but most such soils in-situ and subject to only nominal natural moisture fluctuations do not change volume sufficiently to present problems in most situations. There are, however, some soils whose clay minerals have a capacity for taking water into their crystal lattice with resultant large increases in volume or swell pressure. Some common clay minerals have large capacity for swell, some moderate capacity, and some small. More detail is not pertinent here but can be found in Refs. *1*, *2*, and *8*. The maximum swell to be expected from a particular soil is a matter of the proportion of large and/or moderate swell capacity clay minerals making up the clay fraction, and the proportion of clay in the soil. Swelling of as much as 15% or greater increase in volume from moist (near OMC) to soaked can be encountered. More generally, for highly swelling soils, the volume increase will be in the 7–10% range. Commonly, swelling of 2½ to 3% will not lead to unexpected problems, but swelling even this large should be carefully considered in any critical circumstances. Swell pressures for totally constrained soils can go as high as several thousand pounds per square foot.

It should be apparent that swelling of high-swell soils can be avoided if variation in moisture can be avoided. Normally this is not a feasible means of control but occasionally it can serve the need. Also there are some instances in which substantial swelling can be tolerated. In relation to other than very small structures it is commonly necessary that to be tolerable the swelling be uniformly distributed beneath the structure. This uniformity is rare since it demands uniformity of soil characteristics and distribution, uniformity of water availability and drying conditions, and even uniformity of sun, shadow, and air movement characteristics. The only completely satisfactory treatments of swelling soils are removal and replacement of the complete death of the deposit or to a depth below which there is no moisture variation, or covering such soils with sufficient overburden to sustain the maximum swell pressure without uplift.

Soils will experience their greatest swell on wetting and shrinkage on drying when they have been compacted dry of optimum moisture (see Section 12.9 and Fig. 22.2). Soils compacted on the wet side of optimum have maximum swell a third or less than the same soil compacted dry of optimum (see Ref. *45*, Fig. 9.5). Thus in some instances potential swelling can be reduced by placement of the soil in a too wet condition. The result is a low-strength condition which must be dealt with, but swell reduction can be gained. It must be cautioned that the smaller swelling is due, at least in part, to satisfying the moisture intake of individual particles during placement. Thus the advantage is only satisfactory in the long run if the soils do not later dry and become rewetted.

Concern for swelling needs only to relate to clays and more commonly only to heavier or more plastic clays. Generally swelling clays will be well known in areas where they occur, but testing for swelling will normally be worthwhile.

The foregoing discussion is intended to acquaint the unfamiliar with what to look for and expect in relation to soil swelling, but it is advisable to seek expert guidance for any serious problem or substantial construction being planned which involves swelling soil.

4.2 Sensitive Soils

Some deposits exist, fortunately not commonly, of soils which have a very large proportion of voids, i.e., a high void ratio. These soils have a very open or spongelike structure in which individual grains are in contact, but the grains are not nestled together so as to have multiple contacts. Some such soils were apparently formed by individual grains joining into flocs settling, most often in a saltwater environment. Such formations would have been made in a past geologic time, with no later geologic process working to consolidate the soils. Some sensitive soils are in high rainfall volcanic deposit areas where both the volcanic origin and subsequent leaching of chemical constituents have left the quite porous structure. These soils have fair in-situ strengths, but handling, manipulation of any kind, or loading beyond their capacity to resist will result in dramatic strength loss. When the very open structure is destroyed, the water in the large pore spaces is much greater than more normal soil structures have room for, and the result is a soil slurry with little strength.

While there appear to be some similarities between the sensitivity of some clays and the remolding strength loss of silts (the two phenomena are often considered as one), the clay sensitivity is the result of long geologic processes and once destroyed cannot be restored, while the silts' capacity for remolding strength loss will reestablish in a matter of days.

A significant difference, which on occasion may be of value to know, is that the sensitive clays do not retain their open structure and sensitivity at the natural surface. Normal processes of wetting-drying, heating-cooling, and freezing-thawing apparently destroy the open structure to depths of perhaps two to several feet. Thus, except for unusually heavy, intense, or repetitive loadings, sensitive clays will not present problems except in excavations. The remolding strength loss in silts, on the other hand, is likely to be most severe at the surface and less severe at depths.

The greatest problem that sensitive clays pose is their unexpected inability to support construction equipment and normal construction operations. It is usually necessary to avoid placing loads directly on exposed soils or to distribute the loads widely through "rafts," constructed pads or roadways, or greatly widened tracks on equipment.

4.3 Some Other Unusual Soils

This discussion can only attempt to convey general concepts with some warnings of problems which may require special expertise. There are several other special soils worthy of mention here, but for which descriptive guidance pertinent to individual soils cannot be given in detail.

Caliches or caliche soils can vary from weak limestone to soils with some incorporated lime deposits, and any of these may be referred to locally as caliche. These are formed in areas where severe drying brings lime-water to the surface and evaporation leaves lime deposits. A common problem results from use of standard tests to determine water contents. The oven-drying of the standard test removes bound water or water of hydration as well as moisture merely wetting the soil. The result is an incorrect indication of the behavior to be expected from the soil. Caliches also typically experience significant strength loss upon wetting.

Shales are commonly treated as broken rock or aggregate with the expectation that they will behave as an aggregate or coarse-grained soil. Often such shales will break down on handling, under vehicle traffic, or over longer periods from wetting and drying. The result is a soil behaving as a clay when it was expected that the soil would behave as an aggregate or coarse-grained material.

Laterites and lateritic soils are not common in the United States but are frequently encountered throughout the world. The behavior character, formation processes, etc. of these materials form the basis of many studies and treatises, well beyond the scope of this presentation. Limited knowledge of these materials and false expectations of their performance have led to many unsatisfactory uses. Frequently laterites provide the only available gravel in areas far from a rock source. An often repeated mistake is to define expected behavior using standard tests, without allowing for breakdown of the soft aggregate on handling. In such cases the materials tested are not the same as the materials placed in constructions in the field.

II
Applications

5

Introduction to Part II

5.1 Purpose

The first part of this book presented general concepts of soil behavior to provide contractors and others who are not familiar with such behavior some "feel" for their dealings with soils. This second part presents information more specifically oriented to particular applications, drawing heavily from a prior book, for which one of the authors here was a coauthor. This earlier book, *Construction Guide for Soils and Foundations* by Gordon A. Fletcher and Vernon A. Smoots, continues as a valuable reference to functional applications for heavy foundations. Much of this material has not been carried over into this presentation, consistent with the redirected aims of this book and space limitations.

5.2 Planning Excavation Work

5.2.1 Contract Documents

The contract documents (plans and specifications) direct the contractor what to build. It is generally up to the contractor to select the methods of site preparation and protection, and to be responsible for them. If unanticipated difficulties arise, such as "quick sand," unstable ground, rupture of utilities, or slipping of excavation bracing systems, substantial changes in the method of construction may be necessary. These changes also may change the substructure itself. As a result, the structure may not perform according to the original plans.

While many contracts provide for "supervision of construction" by the architect or engineer, such supervision is usually intermittent and confined to the structure. It is only after serious problems have developed that the engineer becomes involved with the construction methods.

5.2.2 Grading and Site Preparation

The character of site soils and nature of any rock that may be encountered can have significant impact on any construction involving cut or cut-and-fill earth moving. Potential problems with site drainage, support for construction equipment, excavation, erosion, recovery from bad weather, and so on argue the need to learn all one can about site soils, both type and condition.

5.2.3 Shallow Excavations

Shallow excavations can easily be treated too casually. Identifying the character and condition of soils, shallow groundwater, or presence and character of rock frequently helps to avoid undesirable surprises. Site drainage, erosion problems, concern for safety regarding bank caving, rock problems, swelling, and sensitive soils are all potential problems even for shallow excavations.

5.2.4 Deep Excavations

Deep excavations, by their very nature, "ask" for trouble. The flat, uniform, normal ground situation is interrupted by abrupt new changes. Unloading of the ground results in rebound. Undermining of adjacent streets causes lateral movements. Such small movements can break water mains, softening the ground and causing more movement.

The process of excavation, shoring, prestressing of shoring, installation of permanent foundations and walls, and backfilling requires as much planning as a military offensive. These steps must be fitted into a general plan, with alternatives if problems develop and cause time delays in certain phases of the work.

Safe side slope angles for excavations may encroach on streets or adjacent property. Where soils are to be shored, the excavation must be done in steps to provide time and working space to install bracing or tiebacks.

5.2.5 Dewatering

Where the soil information reveals a water level above the subgrade, the decision about dewatering methods should take into account the draining characteristics of the soil and whether pumping may be performed from within or outside the excavation. The effect of dewatering in the area is equally important. Dewatering may cause settlement of nearby structures—particularly if the structures show signs of structural strain, as evidenced by settlement cracks, no matter how thin. Furthermore, buildings on wood piles may be endangered if the lowering of the groundwater will expose the tops of the piles to dry rot. If site dewatering continues for several months, a process of recharging to maintain the water level may be needed.

5.3 Position of the Contractor

As long as the contractor installs the subsurface construction in accordance with the contract documents, his or her position for this part of foundation work imposes no greater responsibility than the superstructure work does. However, the work involved in preparing the site for the permanent substructure becomes the complete responsibility of the contractor for that part of the work. Engineers generally refrain from specifying the methods or design of temporary construction. Frequently, specialist subcontractors are brought in to handle the subsurface work. Several such contractors may be asked to bid. Their inventiveness, special equipment, and knowledge are brought into competition, which may result in the best method and also the lowest price.

Accordingly, the general contractor selects subcontractors to perform the subsurface construction. The subcontractors do their work in accordance with designs for the excavation of safe slopes, sheeting and bracing of banks, and other temporary work. These designs are developed by registered professional engineers working for or retained by the subcontractor. When making designs for site preparation, engineers must consider that the subcontractor is responsible for protecting streets, utilities, and nearby buildings from damage. Occasionally the owner prepares the design for site preparation, in which case he or she retains responsibility for protecting adjacent facilities.

5.4 Available Information

Usually when a contractor has trouble with the soil, rock, or underground, there is information available somewhere which could have helped the contractor to anticipate the problem. Where can this information be found? It is hard work to track it down—and sometimes it is not there. However, it is worth looking for. This book outlines some places to look and gives a checklist to serve as a reminder (see Chapter 7, Section 7.7).

5.5 Method of Presentation

Soils engineering—like most engineering—is partly scientific. The simple and understandable ideas of science can be applied to the solution of everyday problems. The rest (hopefully) is common sense and experience. Engineering also is restrained by economics, building codes, and politics. Therefore, engineering is not infallible and should be questioned. A person who is going to risk dollars in constructing the project should question the soils and the design engineers.

The practice of soils engineering continues to develop rapidly and changes with new findings. The answer given last year may be somewhat different today. Such progress can be confusing. Engineering reports should give as much guidance as possible. A report should also describe enough of the problems (and possible solutions) to suggest to the reader some of his or her own ways to do the job.

Regarding presentation within Part II of this book, the following is true:

1. Many of the chapter titles were selected to match paragraph titles common in bidding documents and specifications. For this reason, there is some duplication between chapters and referral from one chapter to another, and with the general presentations in Part I.
2. The words and methods of description are designed for readability. Complex textbook treatises are omitted.
3. Formulas and mathematics are almost eliminated.
4. References are given to books and other sources which treat this subject in more depth and detail.

6

Available General Information

At most construction sites, some general information is available. This information may already have been researched by the engineer who designed the project. However, the contractor is obligated to research all the available data himself. He must attempt to unearth all history of construction at the site. Review of such data may warn of a problem which is not discussed in the design specifications. Specification writers "cut and paste" copy from previous jobs. They do not always rewrite the specifications to fit exactly the new project.

If there should be a lawsuit or extra costs, it is to the contractor's advantage to demonstrate that he did consult the available data. Some sources of data are presented in the following sections.

6.1 Topographic Maps

The U.S. Geological Survey (USGS) prints maps at various scales. These indicate elevation contours, marshy areas, human-made construction, mines, quarries, borrow pits, and other similar information. These maps are available at most map stores and at offices of the USGS in most major cities. Commonly, topographic maps for particular areas of concern also can be obtained from other sources. Cities, counties, states, the Corps of Engineers, Tennessee Valley Authority, U.S. Bureau of Reclamation, as well as Defense Department agencies—Army, Navy, Air Force—may have coverage of a site of interest.

6.2 Geologic Maps

Geologic maps are published by the USGS. These maps usually are part of a report of the USGS.

Usually, the easiest method of finding geologic data is to contact the nearest regional office of the USGS. Also, one can contact the USGS to see if it has made independent maps of the area desired. Occasionally university libraries contain unpublished geologic mapping. The Geological Society of America, Boulder, Colorado, publishes a "Glacial Map" of the United States.

6.3 Soil Maps

The Soil Conservation Service, a divison of the U.S. Department of Agriculture, has prepared county maps of many of the agricultural counties. While these maps are slanted toward agricultural use, the classifications of various types of soils in a particular area may be helpful.

6.4 Fill and Unsuitable Ground Maps

In many urban areas, maps are prepared by city agencies or sometimes by real estate groups which show portions of the city in which fills have been placed or soil conditions are generally considered to be unsuitable for construction.

For instance, maps have been prepared which show the zone of fill materials placed around the perimeter of Lower Manhattan Island. Also, there is a map of San Francisco showing the original shoreline and the fill placed beyond this shoreline. In Los Angeles, many areas in the harbor area have been filled. These areas are depicted on available maps. Generally, this information can be most readily obtained from the city engineer's office.

6.5 Hillside Maps

In cities having hilly areas, the building department frequently requires more stringent building standards. Therefore, maps are prepared showing hillside areas. Frequently these maps also show areas of previous landslides. Also, some special maps, such as USGS maps, show areas of previous landslides.

6.6 Subsidence Maps

In several areas of the United States, land subsidence has occurred. Usually the subsidence is attributed to overdrafts of the natural groundwater supply or to removal of oil and gas or to underground mines. Maps of subsidence can be obtained for these areas. Generally, the best contact is the city or county engineer. Areas of known subsidence due to removal of oil or water include Long Beach–Terminal Island, California; Houston Ship Canal, Texas; San Jose–Santa Clara, California; and Lake Maracaibo, Venezuela. Subsidence due to coal mining has occurred in eastern Pennsylvania and West Virginia.

6.7 Flooding Maps

Many areas of the United States have been flooded in the past. Some such areas continue to be flooded in each season of heavy rainfall or heavy stream flow due to snow melt. The best source of such information is the office of the local city engineer, county engineer, or U.S. Corps of Engineers. In some counties, a flood control agency is set up, and this is the best source of such information.

6.8 Frost Depth Maps

The depth of freezing is important in establishing depths of foundations and of utility lines. Maps of the United States showing, in general, the depths of frost penetration are available. In addition, required foundation depths for protection against frost are specified by building departments of the respective cities and counties. A typical frost depth map is shown in Fig. 6.1.

6.9 Aerial Photographs

Aerial photographs have been taken of most of the United States. The Soil Conservation Service of the U.S. Department of Agriculture obtains such aerial photography on contract and keeps a library of such photographs. The Forest Service of the U.S. Department of the Interior maintains aerial photographs of many mountain areas. The USGS has broad coverage under its National High Altitude Photography program (NHAP). Micrographic index microfiche or map plots can be obtained from the Earth Resources Observation Systems (EROS) Data Center and photography from the National Cartographic Information Center (NCIC) National Headquarters as follows:

EROS Data Center
U.S. Geological Survey
Sioux Falls, SD 57198
(605) 594-6151

NCIC National Headquarters
National Cartographic Information Center
U.S. Geological Survey
507 National Center
Reston, VA 22092
(703) 860-6045

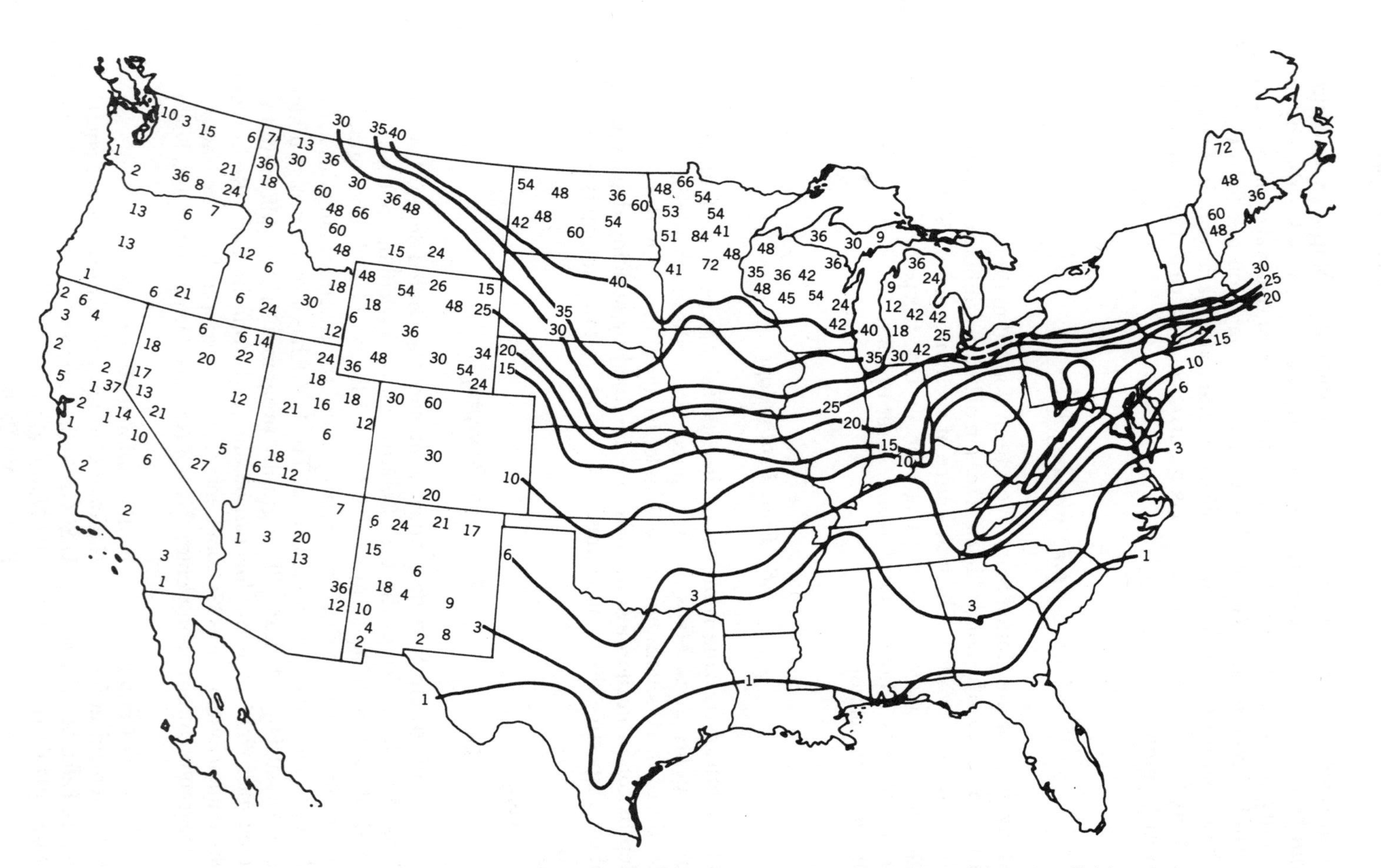

Fig. 6.1 Typical frost depth map (in.) Period 1899–1938. Information collected from unofficial sources. From "Climate and Man," *Yearbook of Agriculture*, U.S. Dept. of Agriculture, 1941.

Information is also available through NCIC regional offices and various federal- and state-affiliated offices. For the address of the nearest office, contact NCIC.

Many aerial survey firms throughout the United States maintain libraries of photographs. In some cases, it is possible to obtain photographs from flights on two, three, or more dates. These sometimes show steps in previous site development. Changes that have occurred on a site over the past 20 or 30 years may show up.

Numerous aerial photographers in or near most cities do photogrammetric work, mainly for engineering firms, architects, and planning departments. Most of them do not have blanket coverage over wide areas. Their activity is usually limited to flying route strips, urban districts, and proposed development tracts.

6.10 Adjacent Construction Data

Information may have been developed for construction of an adjacent structure. This includes boring logs and soil tests. More important, the construction contractor has already pioneered the area. What problems did he have? Is he bidding this job? If not, why not? Did he have trouble with city hall? Take a tour of all nearby buildings. Are they in good shape, or are they cracked or damaged? Look at vacant lots. Is the soil dry and cracked? Is it muddy after a rain?

Check the Street Maintenance Department, the Water Department, and the Sewer Department regarding rebuilding of streets, water line breaks, sewer line breaks, and other maintenance problems. There may be a pattern of frequent repairs around the proposed job site. Check the local library and local newspaper files for old and current photographs. Historical museums sometimes keep files of old road maps and maps showing old shorelines, old bulkhead lines, and old railroad embankments.

A checklist which can serve as a reminder in running a "site check" is presented in Chapter 7, Section 7-7.

7

Available Subsurface Information

On most major construction projects, the owner or design engineer has acquired subsurface soil and rock information to serve as a basis for design. This information generally is reproduced on the drawings, described in the specifications, or made available by reference. It is very desirable to have a prebid conference at which various information can be exchanged, including the available soils information.

On some projects, subsurface information is lacking. It is to the advantage of the contractor to check possible sources of existing available information.

7.1 Investigations by Federal and State Agencies

The Corps of Engineers and other federal agencies drill test borings and obtain other subsurface information on various projects. The state highway department and other state agencies require subsurface information for highway bridges, schoolhouses, and other state structures. Generally, these agencies are willing to let contractors review their report of test drilling.

7.2 Existing Structures

If existing major structures are near the proposed site of construction, information may be obtained regarding the conditions at the site of the existing structure. It is particularly interesting to tour existing structures to examine performance. Generally, the maintenance engineer or building manager is willing to let a contractor tour the buildings. Items to look for include

Cracking of walls due to differential settlements.
Indications of water seepage through basement walls or floor slab.
Tilting of retaining walls.
Settlement of sidewalks through irrigated lawn areas.

These problems can point to soil conditions which should be considered in design.

7.3 Other Contractors

Frequently information regarding construction difficulties can be obtained from other contractors in the area. This might include

High groundwater level, particularly in rainy seasons, or artesian flow of water above general grade.
Difficulty in excavation.
Caving of the walls of excavations.
Difficulty in compacting on-site soil behind basement walls.
Shrinkage and cracking of soils when opened up in excavations.
Unstable soils due to high moisture content or loose soils.

7.4 Engineers Experienced in the Area

The contractor should make a special effort to discuss with the design engineer and the soils engineer the soil and water conditions likely to be found at the site.

Prebid conferences with the engineer and soils engineer should be a *must* on the contractor's checklist of information that may be available for the asking. If other structures have been built in the area, other engineers and soils engineers may have some experience history in the area. Generally, they are willing to make the data available.

7.5 Building Departments

Building plans with accompanying soils reports are filed with the building department for all major buildings. These generally are considered public records and are available. Many building departments permit contractors, or others with a legitimate purpose, to review or copy soil information from nearby projects. In addition, building departments know what kind of foundations generally have been used for nearby buildings.

7.6 Water Agencies

In many parts of the country, water agencies regularly print reports and prepare maps indicating the depth to groundwater at various locations. Such information frequently can be applied to a proposed construction site to indicate groundwater levels and historic fluctuations in the water level.

7.7 Checklist

1. Topographical maps.
2. Geologic maps.
3. Soils maps.
4. Fill and unsuitable ground maps.
5. Hillside maps.
6. Subsidence maps.
7. Flooding maps.
8. Aerial photos.
9. Frost depths.
10. Federal and state agency reports.
11. Inspection of existing structures.
    - Height.
    - Structural frame.
    - Foundation type.
        - Depth.
        - Foundation total load.
        - Bearing pressure.
    - Structural damage.
        - Cracks (should be measured and photographed).
        - Basement floor seepage.
        - Retaining walls tilted.
        - Sidewalk settlement.
        - Street condition (should be measured and photographed).
            - Settlement.
            - Cracking of pavement or of curbs.
            - Pavement repairs.

Distance to adjacent structures close enough to be influenced
  By proposed excavation.
  By proposed dewatering.

12. Other contractors.
    Water level.
    Difficulty in excavation.
    Caving.
    Compaction of fills.
    Shrinkage and cracking of soil.
    Unstable soils or loose soils.
13. Engineers experienced in the area.
14. Local building department.
15. Local water agencies, sewer department, utility companies.
    Utilities in the street.
    Utilities near the site.
    Record of breakage and leaks.

8

Subsurface Exploration

8.1 Surface Examination

Walking over and examing a site can indicate several characteristics. These may include

1. Existence of old landslides. Leaning trees are evidence of past instability of a slope.
2. Existence of sinkholes.
3. Previous fills placed on the site.
4. Previous excavations or cuts.
5. Erosion potential.
6. Previous agricultural use.
7. Cracking of the upper soils, if they are dry. This indicates shrinkage of the soils. Usually such soils are expansive and could be a problem during construction or in the performance of the completed structures.

8.2 Aerial Photographs

Aerial photographs were discussed in Chapter 7. An aerial photograph can be helpful in indicating the points to be examined on the ground. These might include

1. Sharp changes in darkness of tone of the surface soils.
2. Areas of possible outcropping of rock.

3. Areas of heavy growth of plants indicating shallow water table.

8.3 Probings

A ½-in. steel rod, 4 ft long, pointed on one end, and with a handle on the other end, can be helpful for probing the firmness and characteristics of the upper site soils. Tight, clayey soils can be distinguished from loose, sandy soils. The presence of gravel and cobbles can be detected. Also, soft areas and wet areas can be found.

A similar device using about ⅜-in. steel rod with a 1-in. carpenter's bit or auger welded to it makes a handy tool for extracting small disturbed soil samples to limited depth for quick observations.

A much more sophisticated method of probing is the "dutch cone." This device is pushed into the ground with hydraulic jacks which measure resistance at a constant rate of penetration. The cone can be operated separately from a "sleeve" above the cone, so that both end bearing and side friction can be measured.

8.4 Excavations

Excavations and cuts may have been made in the area for highway construction, for general grading in the area, or for installation of underground utilities. Inspection of excavations is important for assessing problems in excavation and stability.

8.5 Test Pits

A Gradall, front-end loader, or bulldozer can excavate several test pits in a day. Such pits, dug to depths of 5 or 10 ft below grade, are an economical way to develop information regarding excavation procedures. They also expose the site soils to these depths for observations and/or sampling.

8.6 Borings

Soil information can be obtained to greater depths in borings than in test pits. In many areas of the country, borings can be drilled using auger rigs or truck-mounted bucket-auger rigs. This type of boring permits examination of the soil removed from the boring on a foot-by-foot basis, and permits a relatively good classification of the various soils encountered. Borings also permit tube-type sampling at the bottom as the hole is advanced. Such sampling is discussed in the next chapter. Several borings can be drilled to depths of 20 to 30 ft in one day using such drilling equipment.

8.6.1 Water Level

During and after drilling, the groundwater level should be carefully measured. For borings drilled with mud, it is necessary to flush out the boring to get satisfactory water level measurements. The borings should be covered and protected, or cased if necessary. Measurements should be taken for a period of time sufficient to indicate the true groundwater level. There may be fluctuations in the water level, a perched groundwater level, or zones of seepage. In many cases, a temporary perched groundwater level will exist for several months after a heavy rainy season, but gradually drain away and not exist during the dry season of the year.

If the groundwater level fluctuates substantially, a long-term series of measurements may be necessary to aid in construction planning (see Chapter 13, Section 13.5 and Chapter 14, Section 14.10).

Water level measurements also may indicate artesian water at depth, and the water pressure or pressure head can be measured.

8.6.2 Depth of Borings

If the excavation will be deep, make sure that the borings were deep enough and that the water level was measured.

For instance, assume an excavation 50 ft deep, borings drilled to 70 ft, all firm soil, firm clay at the bottom of boring, and no indication of water problem. Everything sounds good. But also consider that a sand layer is at 75 ft; it contains water with a hydrostatic pressure such that the water level (WL) would stand at a depth of 10 ft if a boring had been drilled to 75 ft. In that case, when the excavation was opened up, the water uplift pressure down at elevation 75 ft would be 65 ft by 62.4 lb, which equals 4000 psf. The weight of soil overlying the sand is 25 ft by 130 lb, which equals 3250 psf. This calculation says that the bottom of the excavation would uplift, boil, or "blow out." As a rule of thumb, at least one or two of the borings drilled where deep excavations are planned should go twice as deep as the planned excavation. Water levels should be measured in the boring for several days after it is completed.

8.7 Geophysical Methods

Often soils engineering and geology firms have the equipment and qualified people to run geophysical surveys. Such surveys are described in more detail in Chapter 10.

Geophysical explorations can be used to obtain a rough profile of the subsurface materials along lines crossing a site. The method moves quickly, so long lengths of profile can be obtained at a low cost. The information obtained may include

1. Depth of soil underlying the site.
2. Surface of bedrock.
3. Depth of groundwater level.
4. Substantial changes in subsurface conditions from one area to the next.
5. Relative hardness of the rock. (This information can be helpful in estimating the difficulty in excavating rock and whether blasting may be necessary.)

8.8 Geologic Examination

Rock may be exposed at the surface of the site. It should be examined carefully to determine whether it is a boulder or an outcrop of bedrock. This situation should be discussed immediately with the design engineer. Excavations for basements, foundations, or utility trenches could be affected seriously.

The site may be covered with soil and have no sign of rock. However, rock outcrops in nearby areas may be an indication of rock located only a few feet below ground surface.

If it is known that rock underlies the site and will have an influence on the construction, the contractor should find out as much as possible about the rock.

Frequently, bedrock underlying a site will outcrop some distance away. The surface outcrops can be examined to determine the characteristics of the rock underlying the site.

Bedrock at the site may be bedded, jointed, or fractured. These weak zones may be at steep angles. If deep excavations are made into the rock, they are likely to be unstable in steep slopes. Blocks of rock could slide into the excavation. Geologic examination would help to show in which direction instability could be a problem.

Hardness of any bedrock, degree of shattering, and so on, will be of great importance for any local excavation in emplacing drainage or foundation structures or for merely meeting specified grades.

8.9 Offshore

It is more difficult to obtain subsurface information on offshore sites being considered for proposed piers, docks, bridges, oil drilling platforms, new man-made fills, or outfall sewers. Usually such information is obtained by soils engineers experienced in obtaining it. This work generally is obtained prior to bidding. More detailed information regarding offshore exploration can be found in Ref. 9.

Some information can be obtained at offshore locations by bottom sampling. Divers can obtain bottom samples or force tubes into the soil to depths of as much as 5 to 10 ft below mudline.

Deeper cores can be obtained by using a pile hammer to drive pipes into the bottom, or by vibrating pipes or soil samplers into the bottom. Such sampling can obtain penetrations of 10 to 30 ft or more below mudline. Such sampling provides information to be used in estimating problems of stability of the upper soils, in determining pile-supporting capacity, and in searching for suitable materials to be used as hydraulic fill. A photo of a "vibracore" device is shown in Fig. 9.7.

Deeper exploration can be made with drill rigs mounted on barges or on drill ships. The drilling methods must be modified somewhat to account for motion back and forth and up and down.

9

Subsurface Sampling

9.1 Sampling of Soil

9.1.1 Penetration

The standard penetration soil sampler was developed many years ago as a means of obtaining small samples of soils in test borings, *10*.

Samples of the soils encountered are secured by driving a sample spoon or tube into the formation and removing a representative section of soil for visual examination, classification, and preservation (usually in a screw-top glass jar). Since the sample spoon has a wall thickness of ¼ to 5⁄16 in., the soil is remoulded from its natural state. Such samples are called "drive" or "disturbed" to distinguish them from "undisturbed" soil samples.

Based on the theory that the denser or stiffer the soil, the greater the total energy needed to drive the sample spoon a uniform distance with the same driving force, the "standard penetration test" was developed to provide an index of soil density or stiffness.

The sample spoon used for the test has an outside diameter (OD) of 2 in. and an inside diameter (ID) not less than 1⅜ in. It is split longitudinally and held together at the bottom by a shoe shaped as a cutting edge and at the top by a fitting containing a ball check valve. It is a minimum of 24 in. long. The ball valve permits the fluid in the spoon to escape vertically as the sampler is driven and relieves the soil in the tube from water pressure in the drill rods as the sampler is withdrawn. The spoon is shown in Fig. 9.1*a*. The sample spoon is attached to size A drill rods, A shot drill rods, or 1-in. extra-heavy pipe. Other sizes of rods must not be used. A driving head is attached to the top of the drill rods to protect the threads. The

Fig. 9.1 (*a*) Standard penetration soil sampler.

drop weight weighs 140 lb and can be fitted with a rod, or other means, to guide it in its fall (see Fig. 9.1*b*), *10*.

The standard penetration test (symbol N) is the number of blows of a 140-lb weight falling 30 in. required to drive the sample spoon 12 in. into undisturbed soil. Usual practice is to drive the spoon 18 in. and count the blows for each of the 6-in. increments. If the blow count in the first 6 in. is substantially different than the other two, it is discarded. Otherwise, the blow counts for the two increments totaling the lowest number are combined to determine N. Before taking a sample, the spoon should be lowered to rest on the bottom of the bore hole, then tapped a few times to seat it on undisturbed soil before the driving starts. This procedure should also reveal when the spoon has been stopped by an accumulation of gravel which has settled at the bottom of the hole.

To drive the sampler, the drive weight is raised with a length of rope reaved over a single sheave and wrapped several turns on a winch head. After the weight has been raised 30 in., the tension on the winch is fully released so that the weight has virtually a completely free fall.

Because the standard penetration test is largely a manual operation, it is a rough (not precise) measure of soil density or stiffness. A more detailed discussion of its uses and limitations may be found in Ref. *11*.

The standard penetration test may be used to classify the density or stiffness of soil formations as follows:

Granular Soils (N)	Cohesive Soils (N)
0–4 very loose	0–4 soft
4–10 loose	4–8 medium
10–30 medium	8–15 stiff
30–50 dense	15–30 very stiff
50 + very dense	30 + hard

The standard penetration test is less expensive than "undisturbed sampling" and laboratory testing. Therefore, it still is used on many projects and furnishes valuable information.

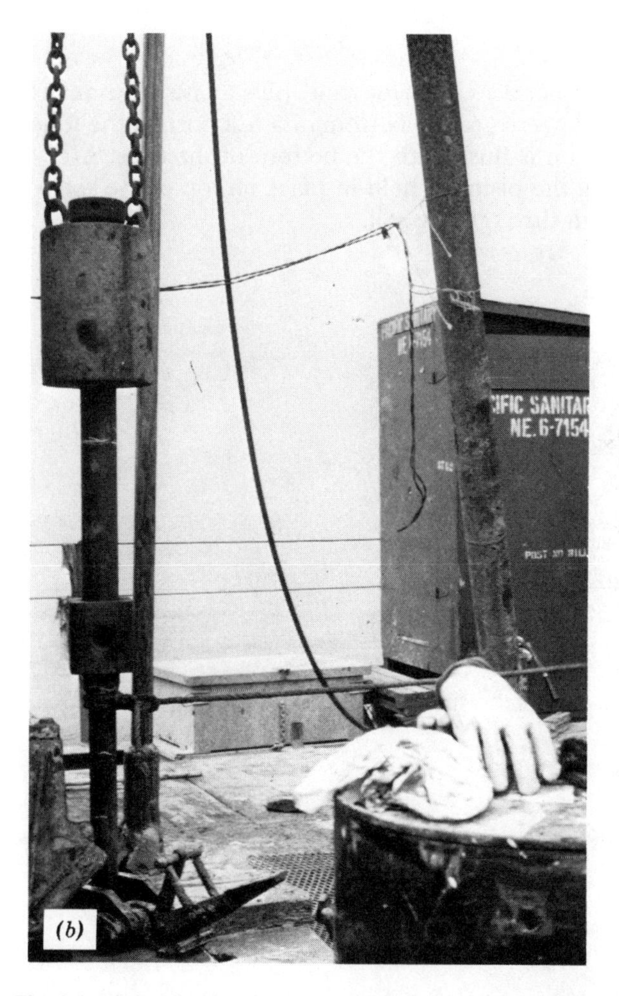

Fig. 9.1 (*b*) Driving mechanism for standard penetration test.

9.1.2 Shelby Tube

The Shelby tube is a thin-walled steel tube which can be fastened to the drill stem of a boring rig and pushed into the soil below the bottom of the boring to obtain soil samples. Usually, these soil samples are relatively undisturbed, since the thin-walled tubing causes little displacement of the soil. Also, it can be pushed or driven into the soil fairly easily. After samples are brought to the surface, the tubes are disconnected from the drill stem; the tubes are capped at each end, or also may be capped with hot paraffin poured into each end. The tubes then are taken to the laboratory for testing, *12*.

The tubes come in several sizes. Popular sizes are 2.0 and 2.5 in.; OD wall thicknesses are from 18 to 11 gauge.

9.1.3 Piston Sampler

The piston sampler uses a Shelby tube, plus a piston inside the tube. When the sampler is lowered to the bottom of a test boring, the tube is pulled up so that the piston is flush with the bottom of the tube. After being seated on the bottom, the piston is held in place on top of the soil while the tube is pushed down through the soil.

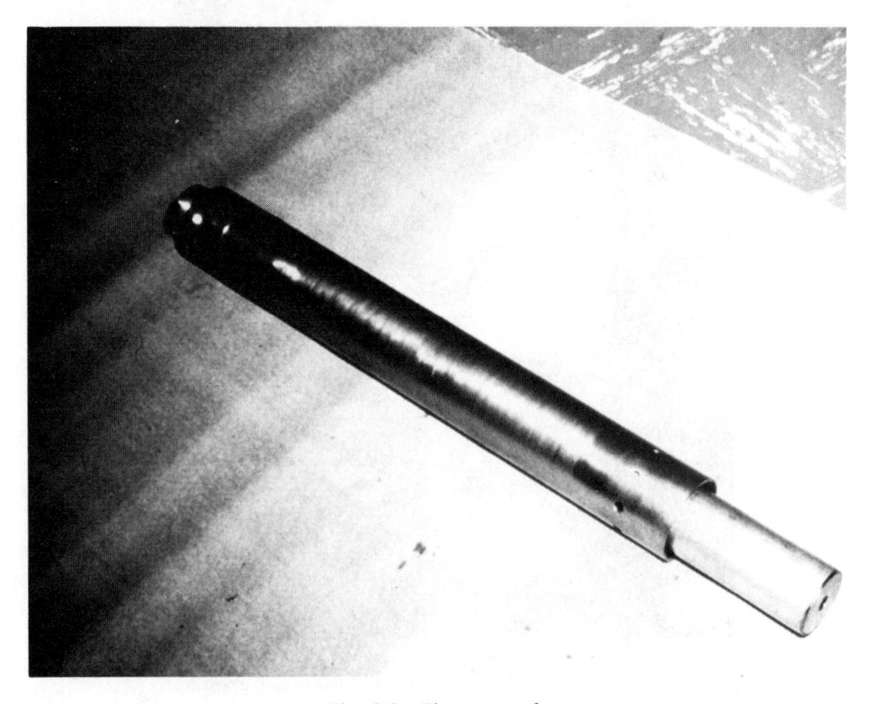

Fig. 9.2 Piston sampler.

It has been found that this sampling method reduces the tendency for the soils to be disturbed during sampling. Also, in withdrawing the sample, the chance of a sample sliding out of the tube is reduced greatly, since a vacuum is produced between the top of the sample and the bottom of the piston (see Figs. 9.2 and 9.3).

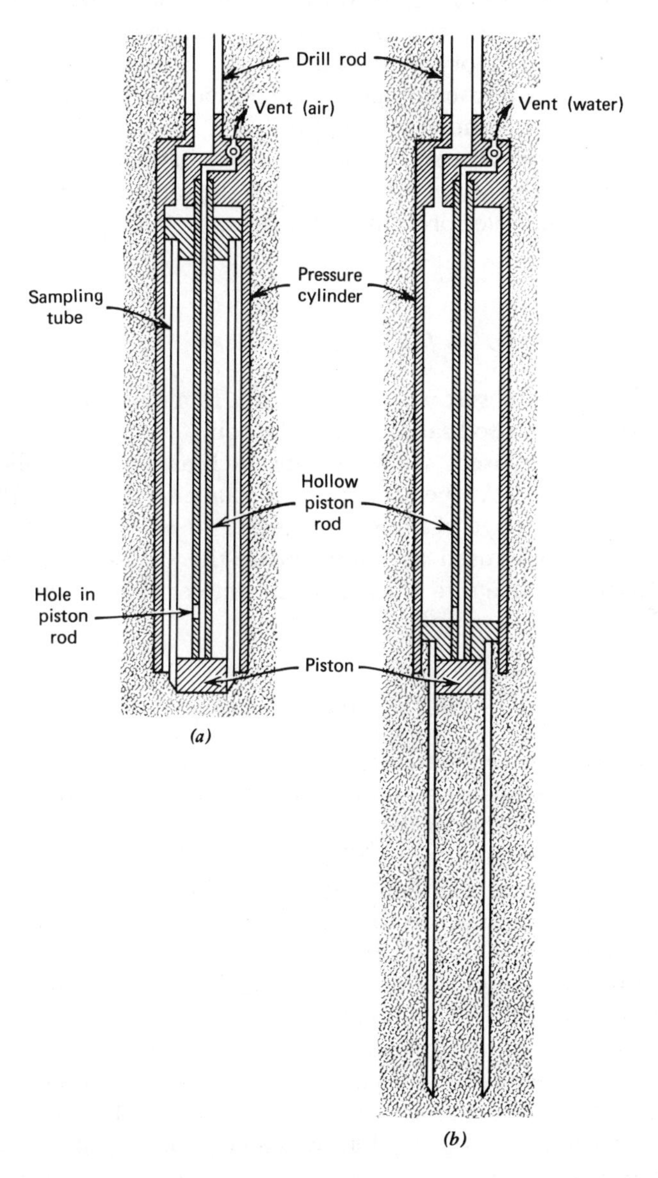

Fig. 9.3 Piston sampler of hydraulically operated type. (*a*) Lowered to bottom of drill hole, drill rod clamped in fixed position at ground surface. (*b*) Sampling tube after being forced into soil by water supplied through drill rod. From Ref. *13*.

9.1.4 Split Barrel

Large diameter split barrel samples are commonly used for obtaining "relatively undisturbed" samples of soil and soft rock. Their convenience in use has made them popular; common sizes are 2½, 2¾, and 3 in. ID. Plastic or metal lining tubes or rings are placed inside the split barrel. After a soil sample is obtained and brought to the surface, the barrel of the sampler can be separated so that the sample can be removed for packaging and shipment to the laboratory. When liner rings are used, they can be placed directly into the testing equipment in the laboratory. A commonly used split barrel sampler is shown in Fig. 9.4.

This particular sampler has catcher leaves in the bit. If the soil sample tends to slide out, the leaves open and catch the sample. Also, the ball valve at the top of the sampler prevents water pressure from pushing the sample out.

9.1.5 Rotary Barrels

In many firm soils or soft rock, it is difficult to obtain suitable samples by pushing or driving a soil sampler into the soil.

Rotary barrels use a sampling tube such as a Shelby tube, which is pushed into the soil. At the same time, a rotary bit digs away the soil just behind the cutting edge of the sample tube.

Two commonly used soil samplers are the Denison sampler and the Pitcher sampler. A diagram of the Pitcher sampler is shown in Fig. 9.5, and Fig. 9.6 shows a photograph of the sampler.

9.1.6 Aids to Sampling

In some cases, soils are extremely difficult to sample without serious disturbance. Also, caving of borings in sandy soils can allow sloughed material to deposit on the bottom of the boring.

A common method for stabilizing soils in a boring is to use driller's mud, which is a mixture of bentonite clay and water. Other chemicals are sometimes added to improve the weight or consistency of the mud. This work is usually done with a churn drill and a rotary rig. It can be done also with a wash boring rig. The mud forms a cake on the wall of the boring and stabilizes the wall. The mud also forms a cake on the exposed surface of sand samples, which stabilizes them during removal from the borings.

In some extreme cases soils have been frozen. Brine tubes are inserted in the ground around the area to be sampled. The frozen soil is cored, then packaged in dry ice during transportation to the laboratory. An alternate is a cryogenic sampler, which uses liquefied nitrogen to freeze the soil in the bit of the sampler.

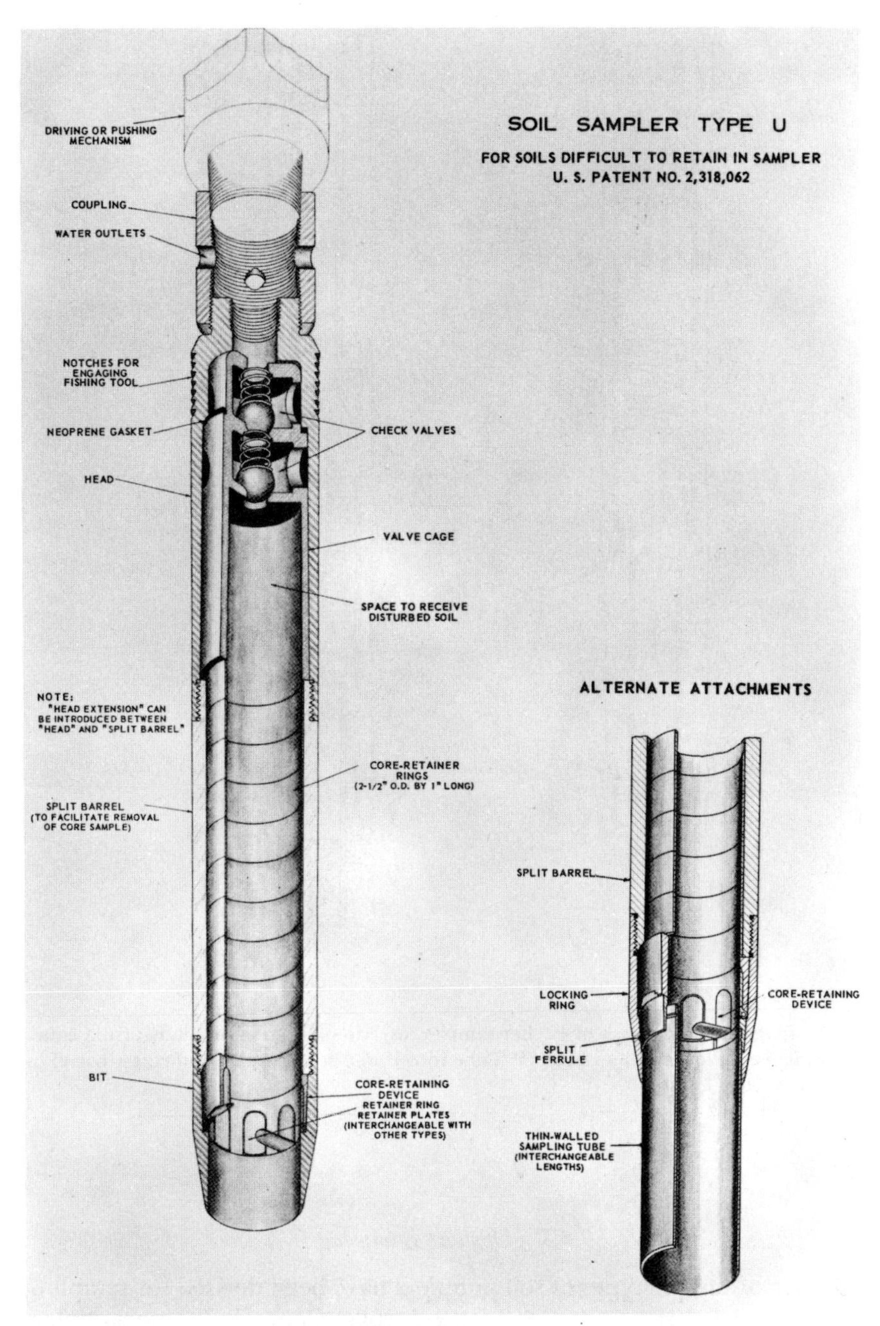
SOIL SAMPLER TYPE U
FOR SOILS DIFFICULT TO RETAIN IN SAMPLER
U. S. PATENT NO. 2,318,062
DRIVING OR PUSHING MECHANISM
COUPLING
WATER OUTLETS
NOTCHES FOR ENGAGING FISHING TOOL
NEOPRENE GASKET
CHECK VALVES
HEAD
VALVE CAGE
SPACE TO RECEIVE DISTURBED SOIL
NOTE:
"HEAD EXTENSION" CAN BE INTRODUCED BETWEEN "HEAD" AND "SPLIT BARREL"
CORE-RETAINER RINGS (2-1/2" O.D. BY 1" LONG)
SPLIT BARREL (TO FACILITATE REMOVAL OF CORE SAMPLE)
BIT
CORE-RETAINING DEVICE
RETAINER RING
RETAINER PLATES
(INTERCHANGEABLE WITH OTHER TYPES)
ALTERNATE ATTACHMENTS
SPLIT BARREL
LOCKING RING
CORE-RETAINING DEVICE
SPLIT FERRULE
THIN-WALLED SAMPLING TUBE (INTERCHANGEABLE LENGTHS)

Fig. 9.4 Split barrel sampler.

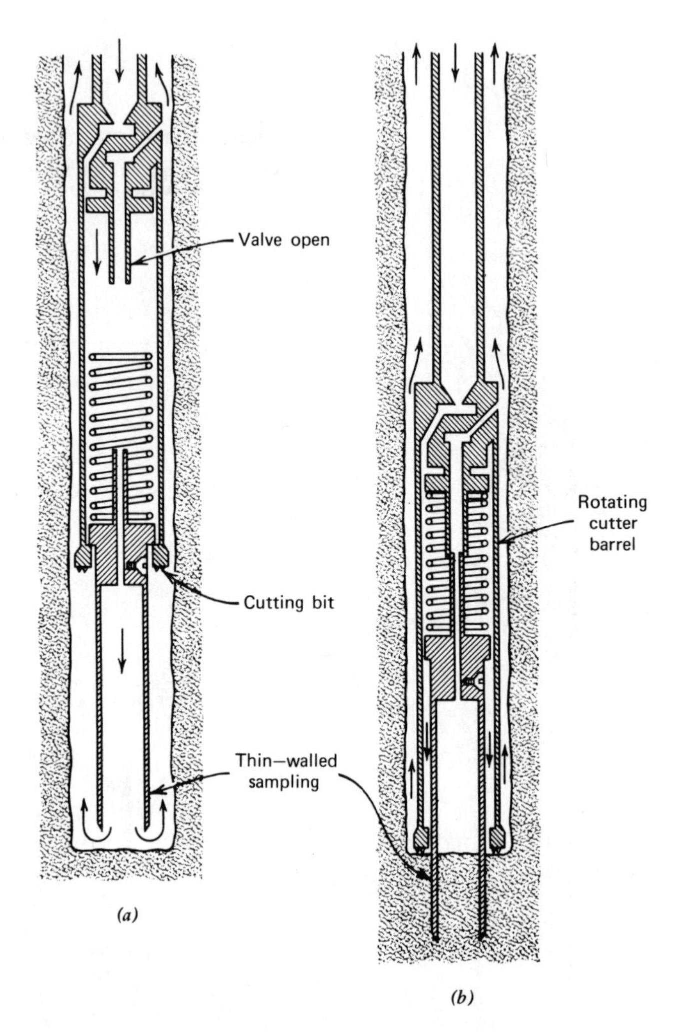

Fig. 9.5 Diagrammatic sketch of Pitcher sampler. (*a*) Sampling tube suspended from cutter barrel while being lowered into hole. (*b*) Tube forced into soft soil ahead of cutter barrel by spring. From Ref. *13*.

9.1.7 Offshore Sampling

A number of unique types of soil samplers have been devised for sampling offshore on lake, river, or ocean bottom soils. Generally, these samplers penetrate to a depth of 2 or 3 ft below the mudline. However, some samplers are long pipes which are either driven or vibrated into the ground. Such a sampling device is shown in Fig. 9.7.

9.1.8 Auger Equipment

Many test borings are drilled using auger equipment. The soils which come to the surface can be gathered, placed in bags, and taken to the laboratory for testing.

The bag samples can be used in tests to determine the compaction characteristics and suitability of the soils for compacted fills.

Soil samplers can be lowered down the hole and then driven into the soil by hand-operated driving equipment, as shown in Fig. 9.8. Also, in holes 24 in. in diameter or larger, it is possible for a person to go down the hole to take soil samples. Short samplers have been devised which can be hammered or jacked sideways into the wall of the boring to take samples (see Ref. *14*).

More recently hollow-core augers, now commonplace, have been employed to allow sampling down through the auger.

9.2 Rock Sampling

9.2.1 Obtaining Samples

Samples of rock generally are obtained by a tube which is set on the surface of the rock. The tube is rotated, and water is flushed through it. At the

Fig. 9.6 Pitcher sampler.

Fig. 9.7 Device for offshore sampling.

bottom of the tube, a cutting edge is built up, which may consist of hardened steel teeth or rows of small industrial diamonds. These are called "single tube" barrels and should be used only in hard, massive rock.

"Double tube" or core barrels are better. They have an inner liner floating free into which the rock core moves during the coring operation. The liner protects the core sample from the turning action of the core itself and from the flow of circulating water. On removal from the hole, the core

Fig. 9.8 Hand-driven sampling device.

barrel is disassembled and the rock core removed. Normally, the cores are packed in wooden boxes.

Core barrels vary from 5 to 20 or 30 ft in length. Commonly used core barrels are 5 or 10 ft in length. When the cores are removed, the total core length is carefully measured and logged. Frequently, the length of core retrieved is on the order of 50 to 80% of the distance cored into the rock. This indicates that some of the rock was soft or fractured and was lost during the coring operations, or that soil in the rock seams was washed away. The length of core recovered versus the distance drilled is called the "core recovery," expressed in percentage. The condition of the drilling equipment and drill bit, as well as the skill of the operator, can affect the core recovery.

It is also helpful to log the pieces of rock core, and in particular to measure the lengths of all of the unbroken pieces of rock core. Such information can be used in a statistical method for comparing rock soundness, called the RQD (Rock Quality Designation) method (see Ref. *15*).

9.2.2 Sedimentary Formations

Some rocks, particularly sedimentary formations, are only moderately hard and can be successfully sampled by soil samplers built of high-strength steel. Core recovery length may be 6 in. to 1 ft. However, this method is much faster than use of core barrels. Sometimes it is helpful in obtaining samples of weathered zones in the rock, which cannot be obtained by the usual rock-coring methods.

9.2.3 Problems

Residual rock formations are produced by weathering, leaching, and chemical action which cause disintegration. However, nodules, slabs, and pieces up to riprap size resist the weathering, remaining hard and unaffected. These solid inclusions are usually scattered throughout the mass, like raisins in a cake. The cores recovered from diamond drilling in residual rock will reveal low percentages of recovery in the disintegrated zones. By contrast, good cores will be secured of the unaffected hard inclusions. Bearing capacities should be assigned by evaluating the strength of the weathered, softer rock, rather than the cores of hard rock recovered. Otherwise, upon exposure by excavation it may be found that the assigned bearing capacity is too high and the job endangered.

10

*Geophysical Exploration**

10.1 General

Unknown subsurface conditions often cause costly delays, and sometimes financial disaster. A contractor courts disaster if he is guessing about subsurface conditions or is relying on subsurface data from a few widely spaced borings. Geophysical exploration is an additional tool that can minimize the guesswork about the soil and bedrock conditions on a project.

The objective of a geophysical exploration as applied to construction work is to detect and locate subsurface soil and rock bodies, to measure certain of their physical properties and dimensions, and to locate other features such as groundwater. The physical properties of soils and rocks measured commonly by geophysical surveying are density, elasticity, magnetic susceptibility, electrical conductivity, or resistivity.

Geophysical methods can be separated into four general classes. Only two are applied to construction projects, which are (*a*) static methods in which the distortions of a static physical field are detected and measured accurately to delineate the features producing them, such as the natural fields produced by geomagnetism, gravity, and thermal gradient or by an artificially applied electrical field; and (*b*) dynamic methods in which signals or energy is sent into the earth, and the returning signals are detected and measured. The dimensions of time and distance always are needed for use of the data.

* This chapter was written by Henry Maxwell and Noel M. Ravneberg of Woodward, Moorhouse and Associates for the first edition. It has been carried, with some deletions, into Part II of this revised work.

10.2 Advantages

The advantages of geophysical techniques over conventional exploration methods are that they are fast, economical, and, especially of advantage in these days of environmental consciousness, nondestructive at the test area; also, because geophysical equipment is readily portable, the building of access roads for heavy equipment is not required. The disadvantage of geophysics is that the results are derived from indirect measurements and therefore geophysical methods cannot completely replace conventional subsurface exploration methods. Geophysics can, however, enhance an exploration program by providing early data to help in the selection of locations to drill borings, and also a profile between boring locations.

10.3 Commonly Used Methods

Two geophysical methods that are well known and the most widely used are electrical resistivity and seismic refraction. Both techniques have long and most successful histories. There have been a few cases where geophysical results have been wrong, causing some engineers and contractors to distrust geophysics. Mistakes were made, especially in the early days of geophysics. However, the methods, equipment, and field techniques have been greatly improved. Geophysics has become more dependable, especially the electrical resistivity and seismic refraction methods.

10.4 Electrical Resistivity

The electrical resistivity technique is based on the ability of a soil or rock to conduct electricity. This ability depends on the salts in the water which occupies the pore spaces in soil or rock. Therefore, the resistance of soils and rocks to the flow of electricity is largely dependent on soil density and moisture. This resistivity gives to the materials characteristic resistances to a current flow. These characteristic resistances, or resistivities, may be used to locate and often to identify subsurface materials and conditions.

The foundation for electrical resistivity methods used in applied geophysics was developed by Wenner in 1915. He introduced a method designed to give a value of apparent resistivity of an earth material below two electrodes. The technique in Wenner's system of electrical resistivity consists of four electrodes equally spaced and in a straight line. An electric current is passed through the ground from a source of direct current applied to the two outer electrodes. Measurement is made of the potential drop between the two inner electrodes (see Fig. 10.1), *16*, *17*.

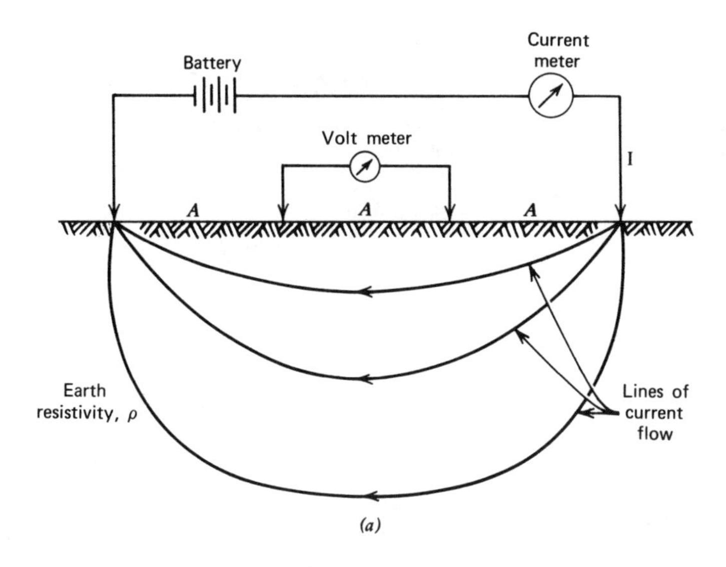

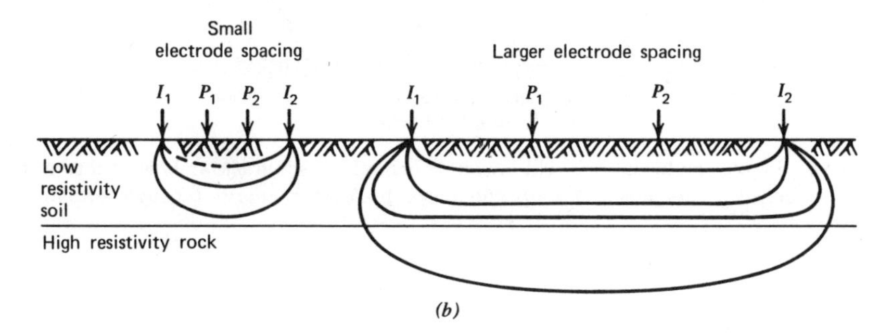

Fig. 10.1 (*a*) Electrical resistivity setup showing lines of current flow. (*b*) The effect of electrode spacing on current distribution and depth of exploration.

10.5 Seismic Refraction

The seismic refraction technique of subsurface investigation consists of creating impact or vibration waves within the ground. This can be done by striking the ground surface with a sledgehammer or drop weight, or by exploding a small explosive charge buried in the ground. The elastic waves created by the impact travel at specific velocities through different materials. The denser the material, the faster the wave moves through the material. Velocities range from less than 1000 ft/sec in loose and dry soils to more than 20,000 ft/sec in a crystalline rock, such as granite. Simply stated, refraction surveys consist of measuring the travel time of an elastic wave between its

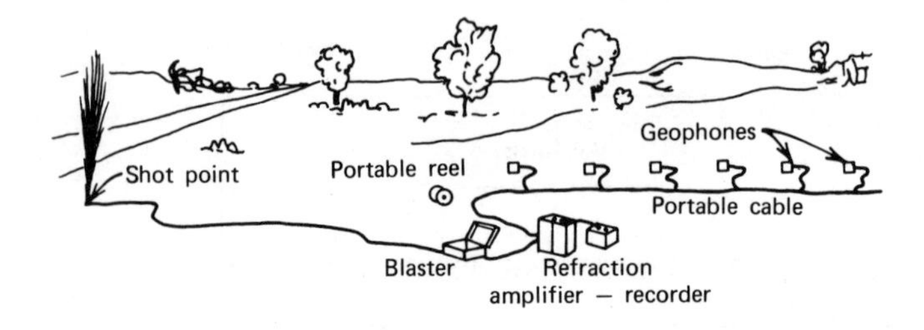

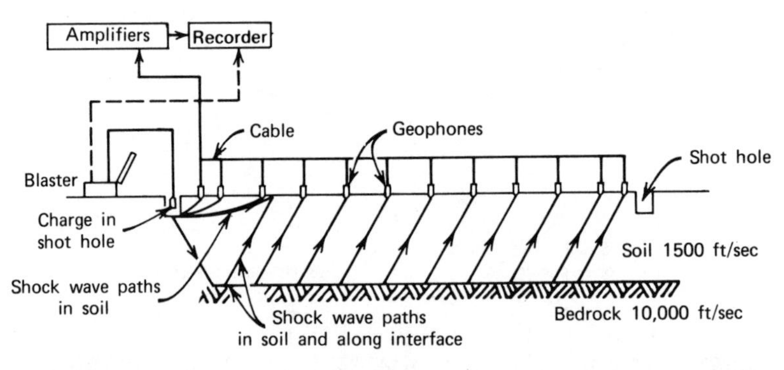

Fig. 10.2 Typical multichannel field layout. Schematic diagram shows how system works. Detectors between shot holes pick up and relay shock wave energy through amplifier to recorder.

starting point and its arrival at a detector, located a known distance from the starting point. This concept is shown in Fig 10.2 and 10.3. The velocity is calculated by dividing the distance traveled by the travel time. Table 10.1 presents the range of velocities typical for earth materials.

In order for the refraction method to work, two conditions must exist, as follows: the velocity must increase with depth, and the various layers through which the refracted wave travels horizontally must have sufficient thickness to transmit the refracted wave.

The seismic refraction method is relatively simple in principle. If subsurface strata consist of homogeneous materials in well-defined horizontally stratified layers, interpretation is easy and straightforward. However, interpretation is seldom simple, because subsurface soils and rocks are usually not homogeneous and occur generally in layers varying in thickness and having complex interface relationships. Interpretation of the seismic data should be done by an experienced person possessing a knowledge of geology and geophysics. Also, the seismic data should be correlated with other existing subsurface information, such as borings.

10.6 General Applications

When, where, and what geophysical methods should be used are questions often asked by an engineer or contractor. The answers depend on the problems, which include the following: How deep is the bedrock? Is the bedrock surface flat, sloping, or irregular? Is the rock rippable? Where is the water table? What is the nature of the soils? Do sands and/or gravels exist in the project area, and if so how much?

The method to use depends on the information required. For depth to rock, identification of materials, and rock rippability, the seismic refraction method gives the best results. Locating the water table and delineating the

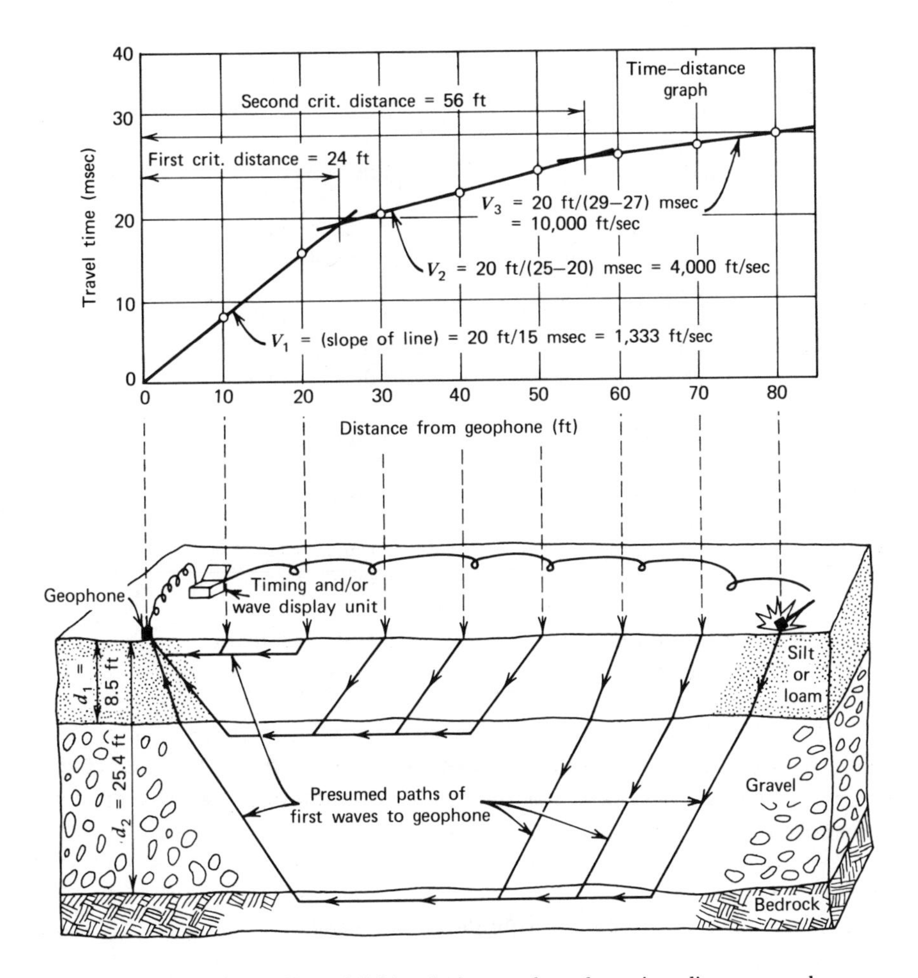

Fig. 10.3 Single channel field technique and resultant time-distance graph.

Table 10.1 Typical Velocities of Earth Materials

Material	Velocity of Sound (ft/sec)
Top soil	
Loose and dry	500–800
Moist loam or silt	900–1,300
Clayey, dry	1,300–2,000
Wet loam	1,500–2,500
Frozen	5,000–6,000
Loose rock, talus	1,200–2,500
Clay	
Dense, wet	3,000–5,000
Gravel	
Mixed with soil	1,000–2,500
Compacted	4,000–6,000
Water-bearing soils	5,000–6,000
Basalt	8,500–13,000
Breccia	
Weathered	3,000–7,000
Solid	6,000–11,000
Chalk	3,000–8,000
Gneiss	
Weathered	2,000–7,000
Solid	6,000–14,000
Granite	
Weathered	2,000–8,000
Solid	8,000–20,000
Greenstone	13,000–18,000
Limestone	
Weathered	3,000–8,000
Solid	8,000–20,000
Quartzite	10,000–20,000
Schist	
Weathered	3,000–6,000
Solid	6,000–11,000
Shale	
Weathered	2,500–5,000
Solid	5,000–13,000

boundaries of sand and gravel deposits are done effectively by electrical resistivity methods. In some instances, a combination of the two methods is necessary.

10.7 Rippability of Rocks

The velocity at which impact waves travel through rock varies with the hardness of the rock. On many jobs, seismic velocities have been compared with the observed hardness of the rock to be excavated and the difficulty encountered in excavating the rock. It has been found that the rippability of rocks can be estimated on the basis of seismic velocities. This is a very helpful application of the seismic method. Figure 10.4 is a rippability chart developed by Caterpillar Tractor Company. A separate chart has been developed for each combination of tractor and ripper.

This method has been used successfully on many excavating projects. However, it must be applied with care. Before committing all equipment and personnel to ripping, be sure the velocity truly represents the condition of the rock. Has the velocity been lowered by open fractures or voids in the rock, or does the velocity represent an average through soft and hard zones? Sometimes a rock gives a high velocity but is still rippable. If possible, make a field test with the proposed ripping equipment.

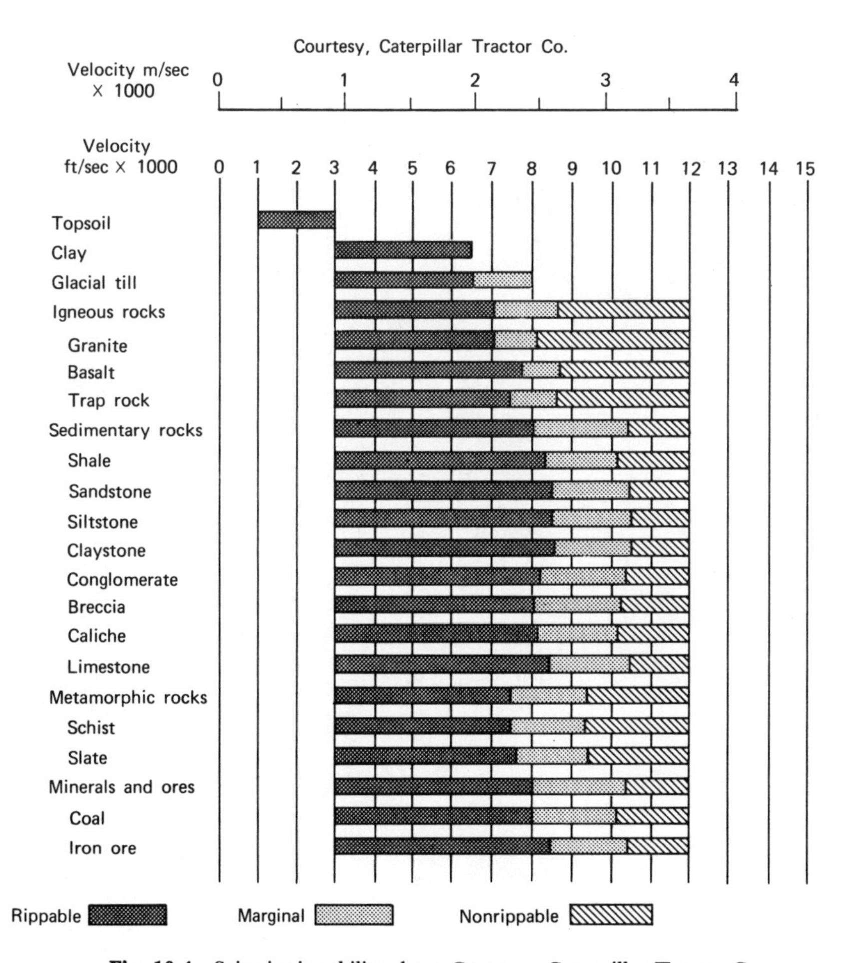

Fig. 10.4 Seismic rippability chart. Courtesy, Caterpillar Tractor Co.

10.7.1 Accuracy

The question of accuracy of geophysical surveying applies most frequently to the depth to bedrock. The depth is usually compared with boring data. It has been found that the shallow refraction seismic method yields results which are generally within a ±10% of the true depth. This accuracy can be as much as 20% or more in error, but more often is less than 5%. The percentage of error diminishes when there are borings with which to correlate overburden velocities and rock depth.

Electrical resistivity is usually the second or third choice if the primary purpose is to find depth to bedrock, because the average margin of error is generally in the range of ±20%. With ideal subsurface conditions (relatively homogeneous strata, mineral and moisture content) and nearby boring information, the accuracy can be improved considerably. However, ideal subsurface conditions for electrical resistivity methods seldom exist.

11

Soil and Rock Characteristics

11.1 Bedrock

Most people consider bedrock a hard material which cannot be excavated without blasting. Generally, the hardest rocks are igneous rocks. These may be granite, basalt, diorite, "traprock," or other similar materials. These rocks resulted from the cooling of molten material underlying the mantle of soil and softer rocks which forms the earth's surface. They are generally excellent for construction purposes.

11.2 Sedimentary Formations

Many areas are covered by sedimentary rocks of various thicknesses. Usually, these rocks are soft, although some are moderately hard or hard. These rock formations are layered, since they are formed of soil particles of sand, silt, or clay laid down in sheets on the bottom of the ocean or lakes. The alternating layers of soil become firm with time, and are classified as sandstone, siltstone, shale, or mudstone.

If the original material consists of seashells and calcareous marine materials, it may be consolidated and altered to form limestone or coral reefs. Limestone may be relatively strong, but is soluble and sometimes develops cavities or sinkholes. Most calcareous formations are loose, easily crushed, and treacherous.

11.3 Metamorphic Rocks

These rocks may originally have been igneous or sedimentary, but they have been altered to form new rocks with different characteristics. Common rocks of this type are gneiss, schist, and slate. Most of these rocks are hard. They have well-developed cleavage planes and tend to flake off in small pieces. For a more detailed discussion of rocks, see Ref. *18*.

11.4 Soil

Soil originated from various rocks. Soil consists of chunks, pieces, fragments, and tiny bits of rock.

Rocks gradually weather, decompose, and soften in place. This decomposed and altered rock changes into soil, which is called residual soil.

If the decomposed rock materials are washed away, usually by rainwater and streams, they are washed down to a lower area where they are deposited in valley bottoms. This soil is classified as alluvial soil.

In some cases the decomposed soil is blown by the wind. The sandy soils form sand dunes. In the midwest United States and other areas of the world, silty soils have been blown great distances in "dust storms." This material frequently builds up in layers hundreds of feet thick. Such soil is called loess. Loess has peculiar characteristics, and working with this kind of soil requires experience (see Section 11.18).

If alluvial soils are carried by streams and rivers out to the ocean and deposited on the floor, they are called marine deposits. Such deposits on the bottoms of lakes are called lacustrine. Marine deposits of sand and silt or clay may become very thick formations. Contrary to some beliefs, sand does not come from the ocean; it is brought to the ocean by rivers and streams. Sometimes uplifting of ocean floors can cause these formations to become mountains and other forms of dry land. These marine sedimentary deposits are the upper soil and rock materials in many areas of the United States. These generally are firm soils—or soft rock.

Some soils were compressed and hardened under the load of the glaciers which previously covered much of the northern half of the United States. They may be called hardpan.

In some western states, volcanoes erupted in the past. Large lava flows covered the surface of the earth for hundreds of square miles. Also, volcanic ash was blown out to form cinder cones. Many smaller mountains are composed of volcanic ash. This material is used for construction of asphalt roadways in Arizona and New Mexico.

Soil, as we see it in the typical construction job, is a mixture of many mineral grains, coming generally from several kinds of rocks. In addition to the mineral grains, the soil contains water, air, or perhaps some gas and organic material, such as roots and humus; it also may contain chemicals.

11.5 Sand

Sand and coarser grained soils are classified in terms of the diameter of the particle sizes. This is indicated by the particle size gradation table (Fig. 12.1).

Sand also can be classified in terms of its grain shape, such as angular, subangular, or round.

Sand generally is considered a favorable construction material, and usually sandy soils are considered favorable from the standpoint of foundation support. Sand is unjustly criticized in the Bible. It becomes a problem in some circumstances, usually due to water. Sand deposits too near the sea or streams may be washed away from under building foundations. Water rising through a sand deposit, due to artesian water flow or other reasons, may create an unstable condition, sometimes called quicksand. On "dry" sites, sand is a good foundation material. It is less likely to develop unexpected bad performance and is good construction material.

Sand does not hold water; water flows easily through it. Any sand which *does* hold water is a mixture of sand and other finer grained soils which plug up the sand. If a sand layer is plugged at the bottom by a silt or clay soil, water can be trapped in the sand. This sometimes is called perched water.

Excavations in sand generally do not stand very well. Dry excavations cave off at slopes of about 1½ horizontal to 1 vertical. Damp sand may temporarily stand steeper, even vertical, for short periods of time. However, sand cut steeper than about 1:1 is very likely to fail within a few days or weeks, and will slide down to a flatter angle, more like 1½ to 1. This is called the angle of repose.

11.6 Silt

Silt commonly is found in flat flood plains or around lakes. It generally has been deposited either by flowing water or by dust storms. It is composed of finely ground-up pieces of rock and is inorganic. Sometimes black organic material is incorrectly called silt.

A dry chunk of silt generally can be broken easily by hand. It is dry and powdery.

Silt holds water reasonably well and is generally soft when wet. A chunk of wet, silty soil, held in the hand and shaken back and forth, flattens out like a pancake and appears "quick." It turns shiny as water comes to the surface.

Silt frequently is found in mixtures with sand or fine sand. Many times a "dirty sand" is a mixture of silt and sand.

Silt generally is not a very good foundation material unless it has been compressed and hardened like a siltstone formation, or has been dried out.

Silt is found in many valleys and streambeds. It is usually loose and wet and generally is easily compressible under low foundation loads, thus causing building settlements.

As a construction material, silt is difficult to use in compacted fill. It is difficult to mix with water. Also, silt tends to be fluffy if dry, or to weave under compaction equipment if slightly too wet.

Some silts are composed of particles which are plate or needle shaped. These silts behave somewhat like clay. Other silts are composed of angular particles, resembling extremely fine sand. These have many of the characteristics of sand. If slow drainage is permitted, their strength characteristics may be similar to fine sand.

11.7 Clay

Clay is composed of rock particles ground extremely fine or reduced by weathering in-situ and nearly always greatly modified in both chemical and crystalline structure by long time trace accumulations from other sources. Clay particles can be flat plates, needle shaped, rounded, or other shapes, but most commonly clay particles have platey structures. A chunk of dry clay is hard and difficult to break by hand. Wet clay can be rolled and molded like modeling clay.

Behavior of clay soil is immensely affected by the vast surface area of the usually platey particles. The individual particles in a cubic inch of clay soil can have millions of square inches of surface area. Not only the capillary, or surface tension, effects of the pore water between particles, but adsorbed or particle surface bound water effects and direct physiochemical particle charge bonds combine to largely determine behavior of clays.

Clay soil commonly contains water in-situ. Natural clay soils have moisture contents in the range of 10 to 50% by weight. If only the effects of surface tension in the soil pores are considered, much can be inferred about clay behavior. It acts like weak glue. If the layer of water becomes very thin, surface tension increases and the glue effect becomes stronger. Chunks of nearly dry clay become very hard.

Although the water surface tension force is small, it becomes large for clay because of the tremendous surface areas, described previously. The small particles are literally held together by the water. When water is withdrawn, by drying, the clay shrinks, cracks, and becomes very hard. This drying process sometimes is called *desiccation*.

Clay soils vary from very soft (and wet) to firm (and relatively dry). Usually, firm clay is a good foundation material. However, the tendency to absorb water causes firm clay to swell. Then it can lift foundations and impose greater soil pressures behind retaining walls (see Section 11.11).

Soft (and wet) clays slowly drain and compress when foundations are placed on them. They are difficult to use as construction materials because they weave and flow under compaction equipment, and are very slow to dry out.

Excavations in clay usually stand well. Firm clays stand in steep, high banks. If the banks are too high or steep, landsliding results (see Chapter 24). Adding water to the clay, thereby reducing the surface tension between the small clay particles, is a primary cause of landsliding.

11.8 Mixtures of Sand, Silt, and Clay

Soils more commonly are a mixture of two or more materials: sand and silt, or silt and clay, or a mixture of all three. The characteristics of the soil therefore are modified. For instance, a sand with small amounts of silt and clay may compact well and provide a very firm resulting soil. Also, the permeability might be quite low, making this material suitable for a reservoir lining.

Soils that contain small grains, medium grains, and large grains are called well graded. Soils with mainly one grain size are called uniformly or poorly graded soils, and soils with sizes missing between larger and smaller grains are also called poorly graded or "gap" graded. Characteristics of some typical soils are shown in Fig. 12.1.

11.9 Mud

Mud is a common though unscientific or nontechnical term. Mud generally is silt or clay or mixture of the two, which contains a large amount of water. Also, muds may contain some organic material. Even sand with some clay or silt can be called "mud" when it is too wet. When muds dry out, they shrink and crack severely.

11.10 Peat

In forests, swamps, thick grass, and other heavy vegetation, dead organic material accumulates on the ground or underwater. Thick beds of decaying organic material may build up. This material may be brown or black and may contain various amounts of soil.

Peat soils are very compressible and provide very poor support for fills or for structures. Also, rotting organic material produces gas such as methane or "swamp gas." If this gas accumulates in manholes or under floor slabs, it can become a hazard. Frequently, such gas kills men working in confined areas.

In colder climates, especially in arctic areas where the decay processes are slow, peats can be common and widespread. Deposits can also be quite thick.

11.11 Adobe

Certain sticky or "fatty" clays are described as adobe or gumbo. These soils absorb water and swell. When they dry, they shrink and crack. These soils occur in many areas of the south central and southwestern states, generally in dry climates. Adobe soils have caused great damage to houses and other structures, and to pavement and sidewalks. The swelling action causes foundations or pavements to move up and down at various seasons of the year. See Section 11.15; Chapter 12, Section 12.11; and Chapter 18, Section 18.13.

11.12 Caliche

Caliche is a soil containing residually deposited chemicals. Caliche occurs in areas of high evaporation rates, typically in desert areas. Evaporation of subsurface water results in chemicals being deposited in the upper layers of soil.

Some caliche soils are extremely hard, like soft limestone. Other caliche materials are more variable and only moderately hard. In some areas where caliche is hard, it is difficult to excavate.

A common error in handling caliche is misdetermination of the soil moisture content. The standard process for moisture content determination in addition to drying a sample drives off water of hydration of the lime and similar components of the caliche and results in an incorrect assessment of the soil moisture. Also, caliches characteristically lose strength on wetting.

11.13 Other Chemicals

Soils may contain various quantities of other chemicals. Some soils are high in sulphides or chlorides, making them "hot soils." Hot soils may cause corrosion of buried utility lines or deterioration of concrete and reinforcing steel. See Chapter 12, Section 12.10 and Chapter 18, Section 18.14.

Calcite (calcium sulphate) is a chemical frequently found in soils. It is slightly soluble in water. There could be concern in a situation such as a dam or embankment subject to continued percolation of large volumes of water. Prolonged leaching may cause failure.

11.14 Water Sensitive Soil

In many desert regions, particularly where flash flooding may have happened in the past, mud flows have occurred resulting in soil of very low density.

This soil generally is hard, because it has dried out in the arid climate. Such soils may have densities on the order of 60 lb/ft^3.

Many residential developments have been thrust into desert areas. In many cases, luxuriant lawns and other landscaping have been supported by imported water.

The low-density soils quickly absorb water. Reduction of surface tension and lubrication permit the soil particles to slide more closely together, with a substantial decrease in the volume of the soil. This sometimes is called collapsing soil. The result is rapid settlement or subsidence of the ground and damage to the structures.

Subsidence of several feet due to collapsing soil has been measured in several locations, such as in the western portions of the San Joaquin Valley in California.

11.15 Expansive Soils

As described in Section 11.11, some soils expand or contract with changes in moisture content. When the volume change is severe it is due to a type of clay called montmorillonite. Soils containing montmorillonite minerals will swell or shrink when water is added or removed.

Severely swelling soils cause many problems and can damage structures of almost any type. In most cases it is advantageous to get expert help in working with soils of high swell potential.

Commercial materials which are largely montmorillonite are called bentonite and have numerous uses in drilling, sealing, and liner applications. The swelling in the presence of water closes pores against seepage flow and tends to stabilize materials (sands) which otherwise would flow and collapse when saturated.

11.16 Frost Sensitive Soils

Nearly all of the northern half of the United States is subjected to sustained cold weather in the winter, sufficient to cause the soils to freeze. The depth of freezing varies from 7 or 8 ft in Maine to 3 or 4 ft in New York and New Jersey to 2 to 3 ft in Kansas City to 1 to 1½ ft in Seattle.

During freezing, the water in the soil expands slightly. More important, if a source of water is available, the soils may draw additional moisture, forming an ice lens or lenses, resulting in expansion of the soils. This uplifting can cause serious damage to structures. In the spring when the soil thaws, the excess ice in the soil turns to water, and the soil turns to mud. Soils which drain easily, such as clean sand or gravel, are not affected. Silt expands most during freezing and turns to mud when it thaws. Clay soils have low permeability and limit the drawing of water. Therefore, expansion is much less than for silt.

11.17 Shock Sensitive Soils

Clean soil containing no binder or other cementing material may be sensitive to shock or to vibration. This applies particularly to clean, loose sand above the capillary fringe zone or below the water table.

Severe shocks, such as may be caused by earthquakes, pile driving, or dynamite explosions, may cause the particles of sand to rearrange and become more compact, resulting in subsidence of the ground.

Loose sands, underwater, which are in the process of densifying, may temporarily lose strength. In this short period, the sand temporarily will not provide support for foundations. This condition is commonly termed liquefaction.

11.18 Blow Sand and Wind-Deposited Silt

These soils have been moved to their present position by wind. They are likely to be moved away from their present position by future winds.

Frequently, new embankments of sand are eroded seriously by wind or rain. A wind-resistant surfacing must be placed over the sand to prevent further wind erosion.

Loess is a wind-deposited silt of very uniform grain size and low natural density. It typically contains vertical tubes or "root holes," and may be somewhat cemented. Vertical cliffs stand well. However, slopes tend to erode and gully, since rainfall softens the soil structure and permits it to flow like sugar when water hits it. If vertical cliffs develop bird nests or gopher holes on top that channelize the water, channel erosion can become severe. This soil is difficult to compact, except with exceptionally close control on the moisture content of the soil. Foundations on loess may settle if the loess becomes saturated.

11.19 Laterite Soils

In tropical areas, the heavy rainfall causes weathering of igneous rocks or leaching of clay soils. This continuous washing may dissolve and remove some of the minerals, resulting in a red-colored soil of low density. These soils may appear firm, and steep cuts can be made into them. However, these soils usually contain a large amount of water. When used as a construction material, these soils become soft and unstable and can be very unsatisfactory.

11.20 Limestone Sinks

Thick deposits of limestone or limey soils occur in the central eastern and southeastern parts of the United States, particularly in Pennsylvania, West Virginia, Kentucky, Tennessee, and Florida. Limestone sinks or sink holes are characteristic of the bedrock formation. Where they occur, they have an effect on the overlying soils.

Limestone is water-soluble to some extent. It may be dissolved slowly by a continuous flow of fresh rainwater, either from the ground surface down through the limestone or by subsurface water coming up to the surface. The water may be slightly acid due to surface organic material or due to acids in the soil. This acidity speeds up the removal of lime. The removal of limestone gradually causes large cavities or "solution channels." The soil overlying the limestone eventually caves in. Sinks generally are round, frequently are full of water, and may contain stands of dense forest growth.

11.21 Hardpan

Hardpan is a soil which has become compact and very hard generally through consolidation under extremely heavy loads, for example, previous glaciers. Hardpan may be developed by other processes, such as natural cementing of a soil layer. Hardpan generally is a good foundation material.

11.22 Dumps

Dumps and sanitary fills are becoming more common in and near many of the major cities. Practice has been to place alternating layers of trash and soil.

Generally, even well-constructed dumps—above water table—consolidate under load. The settlement may continue for many years. In addition, gases may be generated by decomposition of organic material, creating a hazard.

Dumps frequently are converted into parks or golf courses, where subsidence is no problem and the escaping gases pose no problem. However, some old dumps are used for construction of housing or industrial or commercial structures. Generally, there are some problems. Pavements and surface grading have been misaligned due to general subsidence. Local differential settlements affect structures, utilities, and floor slabs.

12

Tests of Soil Samples

In Part I soil types and characteristics were discussed in an attempt to convey an overall awareness of the attributes and behavior to be expected from soil materials as they are encountered in construction. In this chapter the type and character of soils and attributes identified by testing are examined in more detail. Obviously some duplication is to be expected.

12.1 Soil Examination

12.1.1 General Identification

Soil samples usually are described in terms of such basic features as

Color.
Apparent grain size (gravel, sand, or fine grained).
Firm or soft.
Compact or loose.
Wet or dry.
Uniform or variable.
Stratified.
Presence of roots or organic material.
Presence of chemicals such as lime or caliche.

12.1.2 Grain Size

The grain size of the soil is important in identification. Therefore, soil samples are passed through sieves of various sizes, *19*, to determine the proportion of the sample that is gravel, sand, or silt and clay. Figure 12.1 shows some particle-size distribution, or gradation, curves for several soils.

Commonly used sieve sizes go down only to the #200 sieve (nominally 200 openings per in.). The #200 sieve passes particles less than 0.074 mm or 0.003 in. (square opening) in size. The #200 sieve size is the borderline between sand and fines (silt or silt and clay) for a number of extant soil classification systems, *4*, but some—particularly agricultural and geologic systems—adopt the even finer #270 sieve to separate silt from sand.

The array of sizes in soil fines (less than the #200 or the #270, depending on the classification system) is determined by a hydrometer test in which the soil particles are dispersed in water, allowed to settle, and the density of the soil–water fluid checked against time, *20*. Quantities in each size

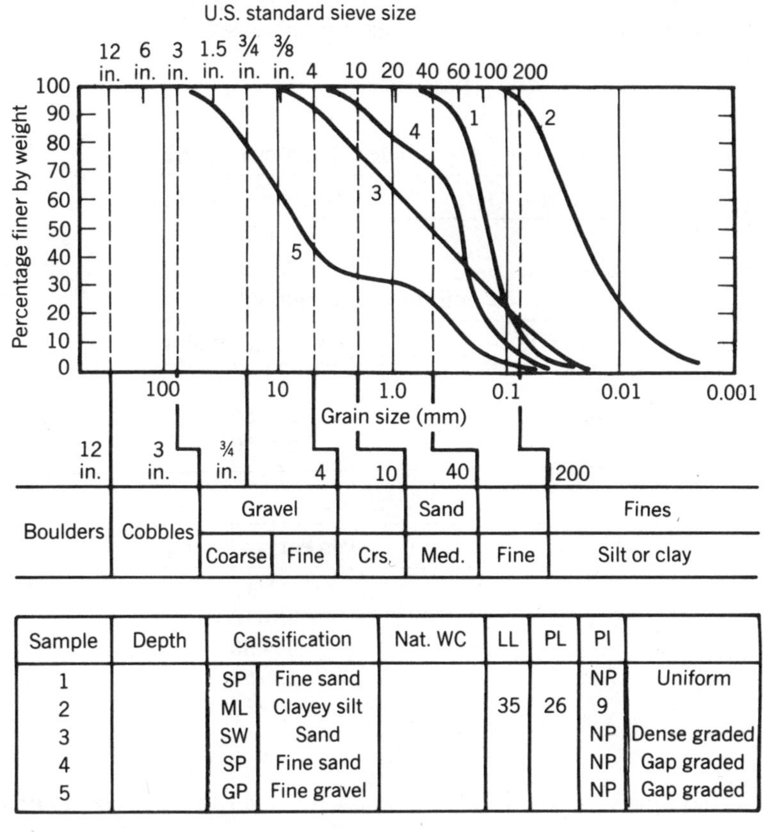

Sample	Depth	Calssification		Nat. WC	LL	PL	PI	
1		SP	Fine sand				NP	Uniform
2		ML	Clayey silt		35	26	9	
3		SW	Sand				NP	Dense graded
4		SP	Fine sand				NP	Gap graded
5		GP	Fine gravel				NP	Gap graded

Fig. 12.1 Particle size gradation curves and characteristics table.

range can be inferred from this test and thus the gradation curve for the fine fraction of a soil can be calculated.

Several important features can be read from the grain size curves.

12.1.2.1 Vertical Curves. These indicate that the soils have been sorted into one particular grain size and can be described as poorly graded, or sometimes as uniformly graded. See curve 1 in Fig. 12.1.

12.1.2.2 Flat Sloping Curves. These indicate a variety of sizes of soil particles, called well graded. See curve 3 in Fig. 12.1.

12.1.2.3 Horizontal Line. This indicates that some particle sizes are completely missing. An example is sandy clay. This sometimes is referred to as skip graded. See curves 4 and 5 in Fig. 12.1, which show a minimum amount of medium sand.

Frequently, materials specified for use on a job are identified by a gradation curve or by a tabulation; examples of specified gradations are given below:

Soil Usually Acceptable for Compacted Fill

Sieve Size	Percentage Passing
3 in.	100
200 mesh	20–30

Select Soil for Compacted Fill

Sieve Size	Percentage Passing
2 in.	100
No. 4	50–85
40 mesh	20–50
200 mesh	5–15

Road Base Course

Sieve Size	Percentage Passing
2 in.	100
1½ in.	90–100
¾ in.	50–90
4 mesh	25–50
200 mesh	3–10

For comparison, a typical gradation for sand to be used in concrete is as follows:

Concrete Sand

Sieve Size	Percentage Passing
⅜ in.	100
4 mesh	90–100
8 mesh	65–90
16 mesh	45–75
30 mesh	30–50
50 mesh	10–22
100 mesh	2–8
200 mesh	0–4

Frequently it is difficult to keep the imported soil materials within the gradation allowances of the specifications. Therefore, the design engineer usually specifies that laboratory tests be made at frequent intervals on the materials to make sure they satisfy the specifications. If the contractor plans to obtain material from a supplier, it is the contractor's responsibility to check the source and make sure the material to be bought actually satisfies the requirements.

12.2 Moisture Content

The moisture content of a soil is a percentage comparing the weight of water to the weight of dry soil. It is determined by the following procedure:

1. Weigh a sample of the soil.
2. Dry the sample in the oven. Reweigh the sample to determine the loss of moisture. The weight of water lost, compared to the weight of dry soil, is called the moisture content. It usually is expressed as a percent. Therefore,

$$\text{Moisture content} = \frac{\text{weight of water}}{\text{weight of dry soil}} \times 100$$

The moisture content generally ranges from about 10 to 15% for sand, 15 to 30% for silt, and 30 to 50% for clay. Some soils, such as bay muds, may have water contents of 100 to 200%. Assuming that a sample of mud has a moisture content of 100%, this indicates that a cubic foot of soil is composed of approximately 45 lb of water and 45 lb of soil grains. By contrast, in desert areas sandy soils may have moisture contents of 5% or less.

12.3 Dry Density

The dry density is the weight of soil particles in a sample. Dry density usually is expressed in terms of a cubic foot of soil.

Most soils have dry densities on the order of 80 to 120 lb/ft^3. However, in the example given above for bay mud, the dry density is only 45 lb/ft^3.

As a general rule, soils having a density of 100 lb/ft^3 or higher are considered fairly good. Sandy soils and well-graded soils generally have higher densities. Silty and clayey soils generally have lower densities.

The soil density depends on two factors: how closely the soil particles are packed together and the specific gravity of the rock or other material of which the individual particles are composed. Most soil solids have a specific gravity on the order of 2.5 to 2.7 Rock weighs on the order of 160 to 170 lb/ft^3 in place.

If a soil has a dry density of 100 lb/ft^3, about two-thirds of the soil is soil particles, and the remainder is air or water. Soils with a dry density in excess of 125 lb/ft^3 are abnormally dense and should be recognized as abnormal. They may contain iron or other heavier minerals. By contrast, a soil containing organic material generally has a lower dry density, frequently in the range of 50 to 70 lb/ft^3.

12.4 Plastic Index

Silty and clayey soils frequently are tested to determine their consistency.

Within a certain range of moisture contents, a given clay or silt can be described as plastic. Below this moisture content, the soil becomes semisolid. At higher moisture contents, the soil changes from plastic to a semiliquid condition. The moisture content of the soil sample can be measured at the limit between plastic and semisolid, and at the limit between plastic and semiliquid. These moisture content numbers are called Atterberg limits, *5*.

The range of moisture contents through which a soil remains plastic is an important characteristic. This range is called the plasticity index, or PI. It is the arithmetic difference between liquid limit and plastic limit. A soil with a plasticity index of 2 has a very narrow range of plasticity. A soil with a PI of 30 has highly plastic characteristics.

Frequently, soils for construction purposes are specified which have a PI below some given amount. Because soils forming the subgrade for roads and highways will become wet sometime in the future, highway departments commonly require that the base course for roadways have a PI less than 6 and in some cases an even lower maximum is specified.

Generally, clayey soils which feel slippery and can easily be molded and rolled out into long strings have a high PI and are unsatisfactory materials for a roadway base.

Occasionally, it is necessary to determine the PI to identify certain soils precisely (see Ref. *21*, p. 30).

12.5 Strength

Several procedures have been devised for measuring the strength of soils in the laboratory. The types of test normally run are as follows:

Direct shear tests—ASTM D-3080, *22*.

Double shear tests.

Torsional shear tests.

Vane shear tests.

Unconfined compression tests—ASTM D-2166, *23*.

Triaxial compression tests—ASTM D-2850, *24*.

The essential features of a direct shear test are shown on Fig. 12.2. A number of shear test devices are shown in Fig. 12.3. As the shearing force gradually causes the sample to fail, the amount of force applied and deflection resulting from that force are plotted on a graph. Such a graph is shown in Fig. 12.4. A point called the yield point, the peak strength point, or the ultimate strength may be picked from this curve and used in later calculations.

If a number of tests are performed using different surcharge pressures, a line connecting these points describes the soil's strength characteristics under various conditions. Such a graph is shown in Fig. 12.5.

A soil which increases in shearing strength with increases in surcharge pressure is generally considered to be granular soil and develops internal friction. Internal friction may be likened to placing two pieces of sandpaper face to face; with no pressure on top of the sheets of sandpaper, they can slide back and forth over each other. However, if moderate pressure is

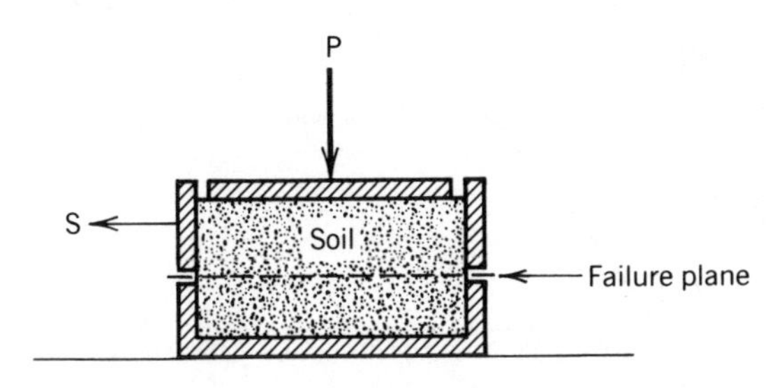

Fig. 12.2 Direct shear test. In this test, the vertical load is usually made equal to natural loading on the soil when it was in the ground.

Fig. 12.3 Direct shear test devices.

placed on the sandpaper, forcing the sheets together, considerable force is required to slide one piece of sandpaper over the other.

The soil may have another characteristic called cohesion. This is a characteristic of clayey and silty soils. These soils have some strength, even with no confining pressure. Cohesion strength is indicated in Fig. 12.5 for a clay soil.

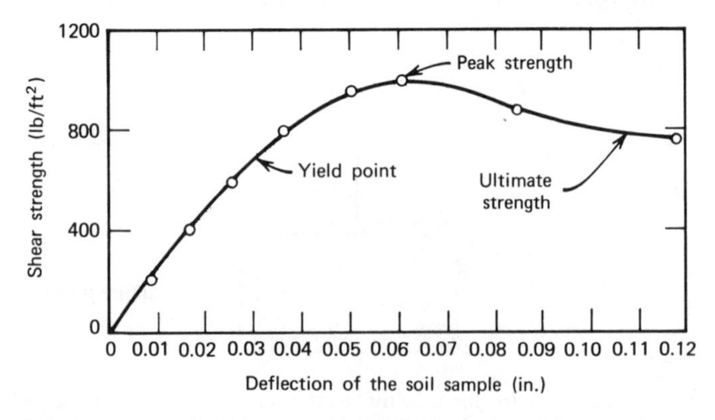

Fig. 12.4 Soil shear strength. Soil strength usually is measured in pounds per square foot (lb/ft^2).

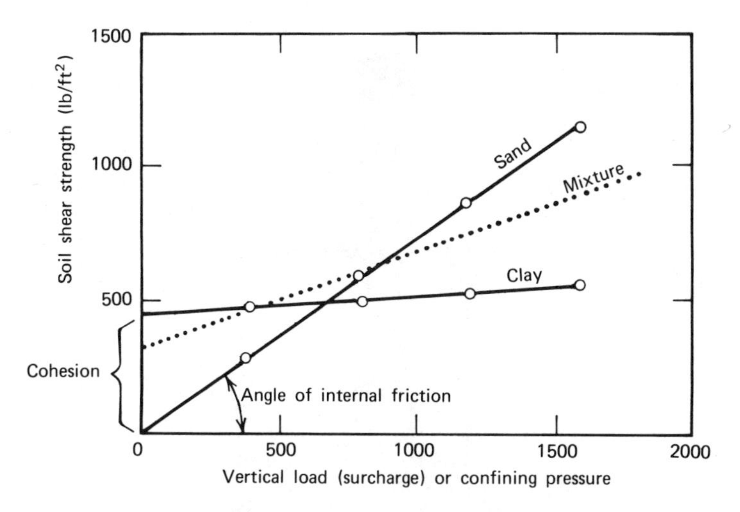

Fig. 12.5 Graph of shear strength vs. loading intensity.

Many soils are a combination of granular soils and cohesive soils, and therefore have both cohesion and internal friction. This is shown in Fig. 12.5. Such drawings are commonly included in soils reports.

During the direct shear test, the soil sample may expand or contract under the confining pressure. This expansion or contraction is measured during the test. If the sample contracts, it indicates a very loose condition and a soil which could become unstable. If the sample expands, it indicates a dense soil structure, which generally would be stable under load, vibration, or saturation.

As discussed in Part I and again above and as shown in Fig. 12.5, sands and other noncohesive materials depend greatly for their shear strength on confining pressure. The laboratory test used to best determine the shear strength at a number of selected confining pressures and thereby to develop the pattern of strength versus confinement is the triaxial test. Figure 12.6

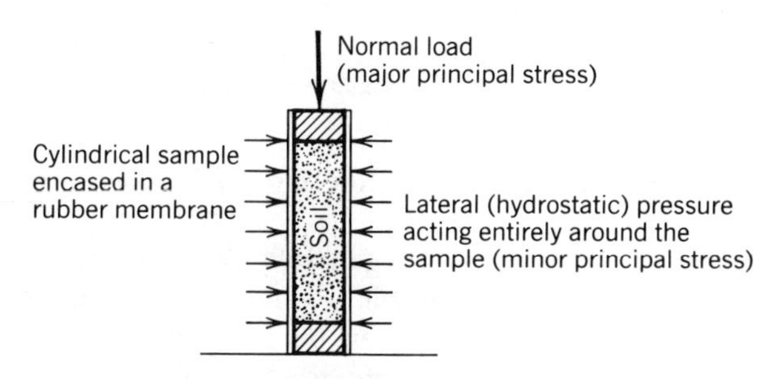

Fig. 12.6 Triaxial compression test.

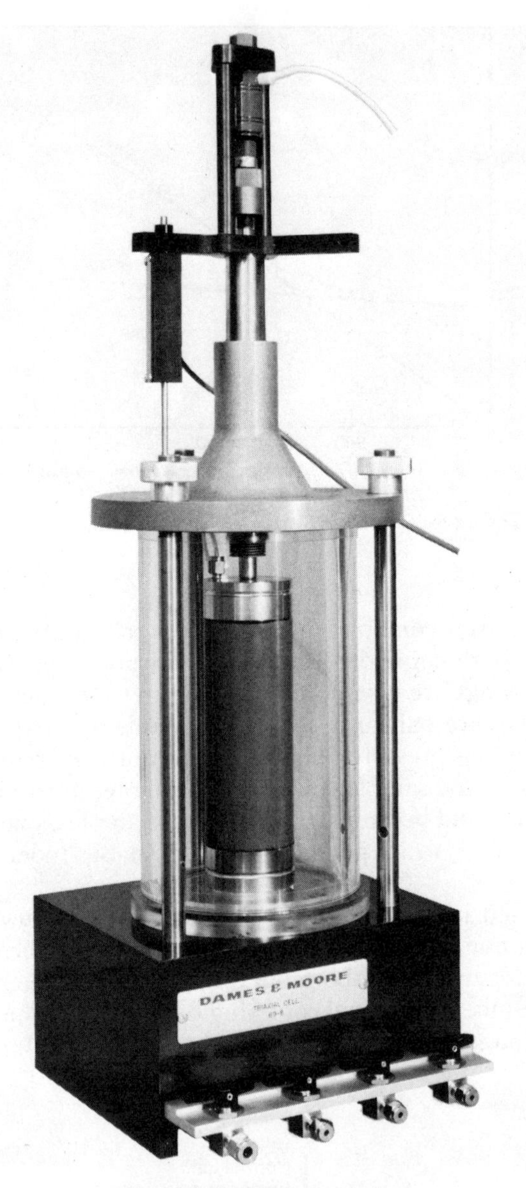

Fig. 12.7 Triaxial test device.

shows a triaxial test schematically, while Fig. 12.7 shows a triaxial test device.

Samples tested in the laboratory generally are undisturbed core samples. However, in many cases it is desirable to check the characteristics of a soil when compacted as a fill. Samples can be compacted to various densities and similar strength tests can be performed (see Chapter 22).

12.6 Consolidation

Consolidation tests are performed to estimate compression or consolidation of soil layers under load. In this way, the settlements of foundations can be estimated. Also, settlements due to placing of earth embankments or heavy loads on the soil can be estimated. The methods of estimating settlements are described in many textbooks and are not repeated here (see Ref. *25*, Chapter 12). A typical consolidation test machine is shown in Fig. 12.8.

As a general rule, consolidation tests are performed on samples 1 in. in height. Consolidation under laboratory loads varies from fractions of a percent up to several percentage. For an example, let us assume that in a

Fig. 12.8 Typical consolidation test machine.

particular case a load on the soil sample of 4000 lb/ft^2 results in a consolidation of 1% of the sample height. Assume that under a proposed foundation there is a layer of this soil approximately 8 ft thick. If the average increase in stress in this soil layer is 4000 lb/ft^2, the soil layer will consolidate 1% of its thickness, or approximately 1 in. The resulting foundation settlement therefore would be approximately 1 in. This example is very rough, but demonstrates the procedure for calculating settlements (see Chapter 19).

The consolidation test can also be checked to determine the speed at which settlement occurs. For sandy or free-draining soils, consolidation occurs quickly. For clayey soils, which are very slow draining, consolidation takes a long time (see Fig. 12.9). Calculations can be made of the speed of settlement to be expected for foundations placed on various soils. For foundations on sandy soils, settlements may occur as the building loads are applied, and be completed within a few months after the building is completed. By contrast, foundations on clay soils may settle over a period of several years. The term "consolidation" is different from "compaction" used elsewhere in the book. Consolidation usually is considered to be a process in which water is squeezed out of the soil, allowing the soil grains to move a little closer together. Consolidation is vertical downward movement of the soil surface. By contrast, compaction usually is considered to be mechanical

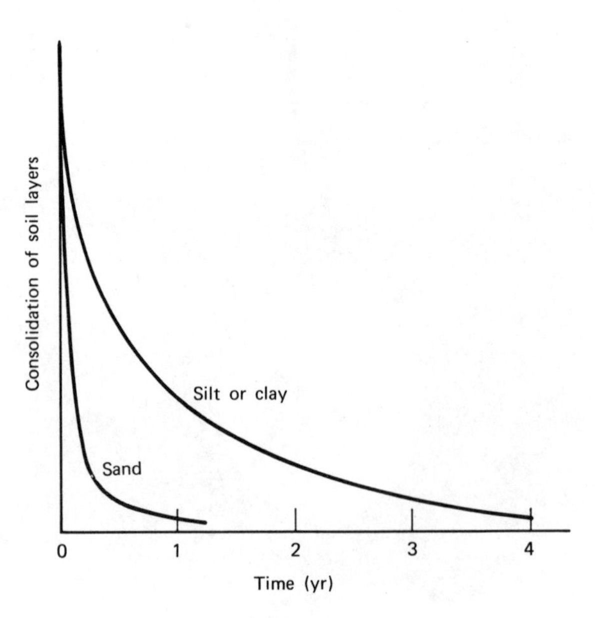

Fig. 12.9 Graph of consolidation vs. time.

rearrangement and densification of soil particles, performed rapidly by rollers, tampers, or other machinery.

12.7 Unconfined Compression Test

The purpose of this test is to determine the ultimate compressive strength and ultimate shear strength of cohesive soils. Typically, undisturbed samples recovered in Shelby tubes are trimmed to a length twice the diameter and subjected to axial compression. The results are plotted as a stress-strain curve from which the peak load or ultimate compression strength (Q_u) is selected.

Assuming that the sample fails in shear along a 45° plane from the axis, it can be shown that the strength in shear is one half the compressive strength.

The test yields valuable information and is widely used. A schematic diagram of an unconfined compression test is shown in Fig. 12.10. Details of the procedure and apparatus can be found in textbooks on soil mechanics and in Refs. *13* and *23*.

12.8 Percolation

Percolation tests are run on soil samples to measure the speed with which water can drain through the soil. During the test, the sample usually is confined by pressure equal to the overburden soil pressure at the depth from which the sample was obtained (see Fig. 12.11). Typical percolation rates for various soils are listed below. Percolatiön rates are used to select dewatering methods for excavations below water level, as well as for other purposes.

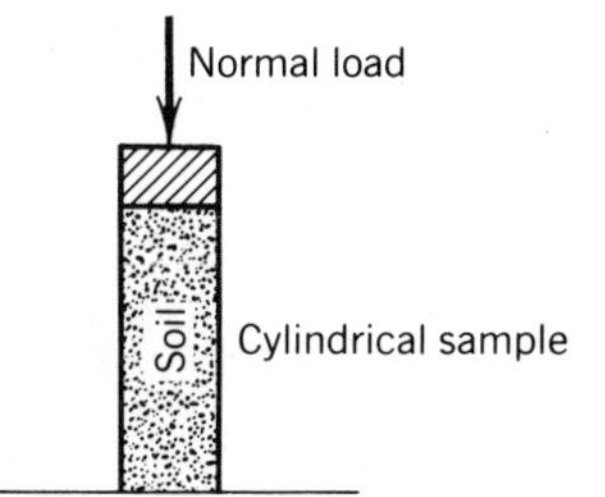

Fig. 12.10 Unconfined compression test.

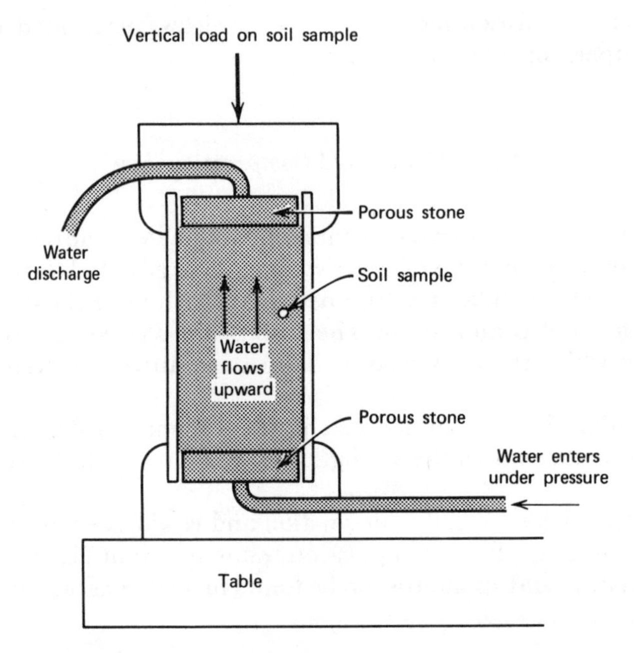

Fig. 12.11 Percolation test device. Rate of flow of water through the system is measured.

Typical Percolation Rates

Soil Type	cm/sec	ft/day
Gravel	10	30,000
Coarse sand	1	3,000
Medium sand	0.1	300
Fine sand	0.01	30
Very fine sand	0.001	3
Silt	0.0001	0.3
Silty clay	0.000001	0.003
Plastic clay	0.00000001	0.00003

For more information, see Ref. *26*, Chapter 2. Percolation rates can be greatly reduced by compaction of soil. For instance, reservoir linings frequently are made of compacted soil. In one case, a silty sand after compaction worked well as the lining of a reservoir.

12.9 Compaction

Compaction tests are run on soils encountered or considered for use in construction where soil density is of concern. Commonly this is in compacted fills, but will extend to subbase and base materials in pavement structures, and can relate to existing soils beneath structures.

The laboratory tests are intended to duplicate the rolling action of compactors in the field. The original Proctor test, using a 5.5-lb hammer dropping 12 in. to compact soil into a mold in three layers, represented the energy level of rollers in use at that time.

With higher demands for roadway and airport pavements, heavier rollers have been developed. Laboratory test procedures have changed to use heavier hammers and more energy in compacting the soil. A commonly used test employs a hammer weighing 10 lb, dropping 18 in., and compacting the soil in five layers.

The original test by R. R. Proctor has become a standard test (see AASHTO T-99 or ASTM D-698) *27*, and the compaction involved is referred to as "standard Proctor" or "standard AASHO" effort. The standard test uses a 4-in. diameter mold having a $\frac{1}{30}$-ft^3 volume, 5.5-lb hammer, 1-ft drop, and 25 blows per layer; the standard effort results in 12,375 ft-lb/ft^3 of compaction energy E.

$$E = \frac{5.5 \text{ lb} \times 1 \text{ ft (drop)} \times 25 \text{ blows per layer} \times 3 \text{ layers}}{\frac{1}{30}\text{-ft}^3 \text{ volume}}$$

$$= 12{,}375 \frac{\text{ft-lb}}{\text{ft}^3}$$

The standard Proctor or standard AASHO test can also be run using a 6-in. diameter mold having a $\frac{3}{40}$-ft^3 volume, 56 blows per layer, on 3 layers.

$$E = \frac{5.5 \text{ lb} \times 1\text{ft (drop)} \times 56 \text{ blows per layer} \times 3 \text{ layers}}{\frac{3}{40}\text{-ft}^3 \text{ volume}}$$

$$= 12{,}320 \frac{\text{ft-lb}}{\text{ft}^3}$$

The later test devised for the higher compaction demands of heavy airfield pavements has also become a "modified" standard test (see AASHTO T-180 or ASTM D-1557, *28*), and the compaction involved is referred to as "modified Proctor" or "modified AASHO" effort. The standard test uses a 6-in. diameter mold having a $\frac{3}{40}$-ft^3 volume, 10-lb hammer, 1.5-ft drop, and 56 blows per layer. The "modified" effort results in 56,000 ft.-lb/ft^3 of compaction energy E.

$$E = \frac{10 \text{ lb} \times 1.5 \text{ ft (drop)} \times 56 \text{ blows per layer} \times 5 \text{ layers}}{\frac{3}{40}\text{-ft}^3 \text{ volume}}$$

$$= 56{,}000 \frac{\text{ft-lb}}{\text{ft}^3}$$

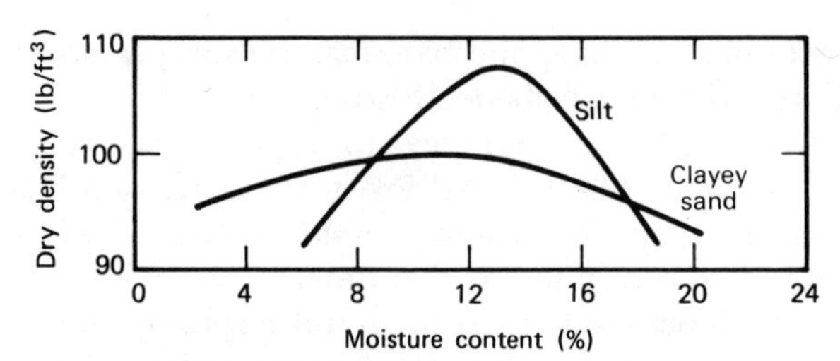

Fig. 12.12 Typical compaction curves.

The modified Proctor or modified AASHO test can also be run using a 4-in. diameter mold having a $\frac{1}{30}$-ft^3 volume, 25 blows per layer, on 5 layers.

$$E = \frac{10 \text{ lb} \times 1.5 \text{ ft (drop)} \times 25 \text{ blows per layer} \times 5 \text{ layers}}{\frac{1}{30}\text{-ft}^3 \text{ volume}}$$

$$= 56{,}250 \frac{\text{ft-lb}}{\text{ft}^3}$$

In the modified test using the 6-in. diameter mold, the "standard" effort can be approximated by applying only 12 blows per layer.

$$E = \frac{10 \text{ lb} \times 1.5 \text{ ft (drop)} \times 12 \text{ blows per layer} \times 5 \text{ layers}}{\frac{3}{40}\text{-ft}^3 \text{ volume}}$$

$$= 12{,}000 \frac{\text{ft-lb}}{\text{ft}^3}$$

In any case, the test involves compacting specimens of soil at each of several moisture contents using the selected method, equipment, and effort. The resulting (dry) densities are plotted versus the moisture contents used. With proper selection of compaction moisture contents the plotted points will describe a curve rising to a maximum and falling away. The maximum point of this curve defines the "maximum density" and the "optimum moisture content" for the soil subject to the applied compaction effort.

A typical set of compaction curves is shown in Fig. 12.12. Note that one curve is fairly flat; the other is steep. The steep curve indicates trouble if the soils are too dry or too wet. Compaction of this soil is not possible if the moisture content is too low or too high (see Fig. 22.2).

12.10 Chemical Tests

Chemical tests may be performed on soils to ascertain whether the soil is acid, alkaline, or neutral and to determine if it contains sulphides or chlorides or other chemicals which might cause deterioration of concrete or steel foundations or pipelines placed in the soil. Chemical tests also may be needed on samples of water and samples of filter materials for the design of drains or wells. It is common to test the pH of soil samples first. If the soil is about neutral (pH = 7), usually no other tests are run. If the pH is high or low, indicating alkaline or acid conditions, additional tests usually are run to measure the sodium, chlorides, or sulfates in the soil. This may indicate the need for special protection for concrete or steel structures placed in the ground (see Chapter 18, Section 18.14).

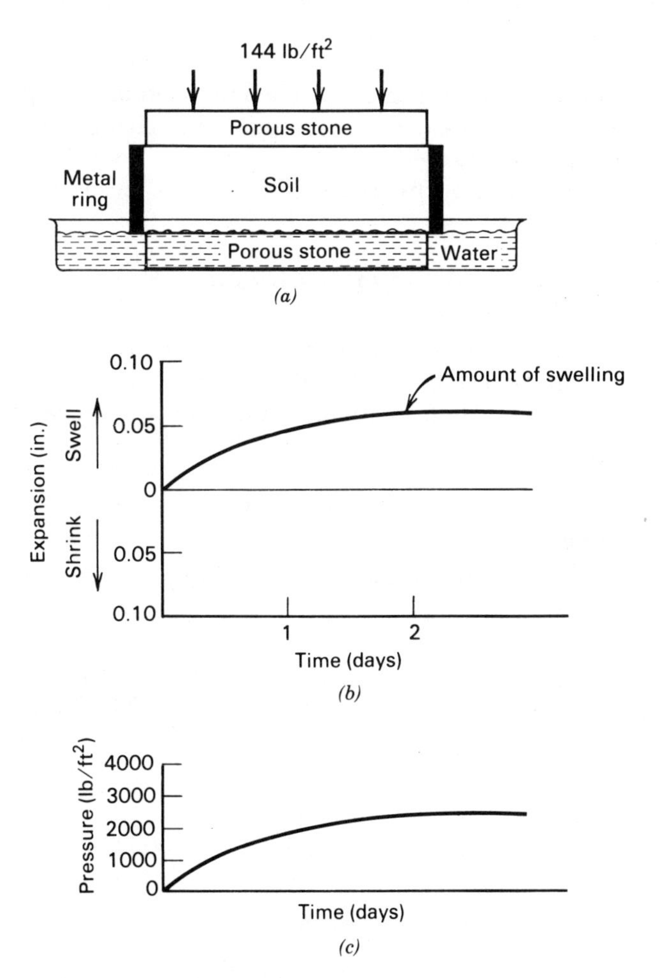

Fig. 12.13 Expansion test. (*a*) Soil sample is placed under a confining pressure of 144 lb/ft^2. Water is added to the soil sample. Sample swells, and the swelling is measured in inches (*b*). The test can also be run to measure the pressure required to prevent expansion. The test data might look as shown in (*c*).

12.11 Expansion Tests

Some clayey soils and shales change volume with changes in moisture content. When these soils dry, they shrink and crack. When they become wet, they soak up water and expand (see Chapter 11, Section 11.15).

Common names for such soils are adobe, gumbo, and bentonite.

Samples of soil can be tested to determine the likelihood that a soil will expand or contract. The sample is placed in a machine similar to that for the consolidation test. The sample is permitted to soak up water and expand. The expansion is measured. Usually, a surcharge pressure (of 60, 100, 144, or perhaps 500 lb/ft^2) is applied during testing. In some cases, in which a predetermined bearing pressure is to be used on foundations placed on the soil, a higher surcharge pressure is used during the expansion test. Expansion of the sample is measured. Then the sample is placed in an oven and dried out. The shrinkage is measured. If the change in volume from saturated to oven-dry exceeds 4%, it is generally considered expansive. Change in volume of over 10% is considered very expansive (see Fig. 12.13).

12.12 Rock Tests

Samples of rock also are tested in the laboratory. Tests are run on rock samples from the area to be excavated or built on, and also on rock to be imported as a construction material. Some tests are similar to soil tests. Different tests include

Abrasion tests, which measure wear and breaking of pieces of rock in a rotating drum.

Freeze-thaw tests, which measure rock splitting or deterioration.

Chemical tests, which measure lack of resistance to chemicals.

Laboratory tests of soil and rock are discussed in more detail in publications of the American Society for Testing Materials (ASTM) (see Ref. *29*).

13

The Soils Report

Soils reports frequently are available at the time of bidding of jobs. Reports can be very helpful in gaining an understanding of the site conditions, construction planning, water problems, soils that may be difficult to work with, and the bad effects or delays that may result from bad weather.

13.1 Types of Reports

Reports may be made at various stages in planning or design of a project. Sometimes, reports during early feasibility studies are limited in scope, and present only generalized opinions based on limited data. Usually such reports are marked "preliminary" or "feasibility" report.

During the course of design, a more detailed report is made. This report usually involves drilling test borings or digging test pits, spending some money for undisturbed soil samples for laboratory testing, performing soil bearing tests, or even driving test piles or performing pile load tests to provide information needed for the design of foundations.

Reports also may be needed on special problems in excavation, foundation construction, or the suitability of compacted fills, backfills, or other materials. Sometimes reports are needed for planning a system of site dewatering or a system of shoring and bracing.

Occasionally, completion reports concerning the soils and foundation construction are prepared for use of the owner.

Should a foundation failure, settlement, or other unfortunate occurrence develop after construction has been completed, reports generally are obtained with opinions regarding the cause of damage and necessary repair work.

13.2 Usual Contents of Reports

A report of a foundation investigation usually includes the following items:

Scope or outline of the work.
Proposed construction site investigation method.
Site conditions found.
Results of laboratory tests or of field tests.
Results of geological or other examination.
Site history or problems characteristic of the area.
Recommended type of foundations.
Recommended design values for foundations. This item would include bearing values for spread foundations or the supporting capacities of piles. This should include the depth for bearing of foundations and the soil type at that depth.
Foundation settlements. Frequently estimates are made of the likely settlements of spread foundations of various sizes and depths, and estimates of settlement are also made for individual or groups of piles.

13.3 Log of Borings

There is no standard form of log of borings. Each foundation company or other organization preparing boring logs has its own method of preparing them. However, a typical log of borings is shown in Fig. 13.1.

Information which should be contained on a boring log includes

1. Boring number.
2. Location of boring or coordinates.
3. Ground surface elevation.
4. Depth of boring.
5. Date drilled.
6. Name of drilling inspector, drilling contractor, and driller on the job.
7. Depth of casing, if used.
8. Type of drilling equipment, such as auger drill, churn drill, rotary drill, or wash boring. Also, whether the hole was drilled dry, with water, with driller's mud, or was cleaned out by compressed air.
9. Whether caving occurred in the boring, and if so at what depths, and the severity of the caving.
10. If a boring was abandoned, the cause for abandonment and the locations of offset borings.

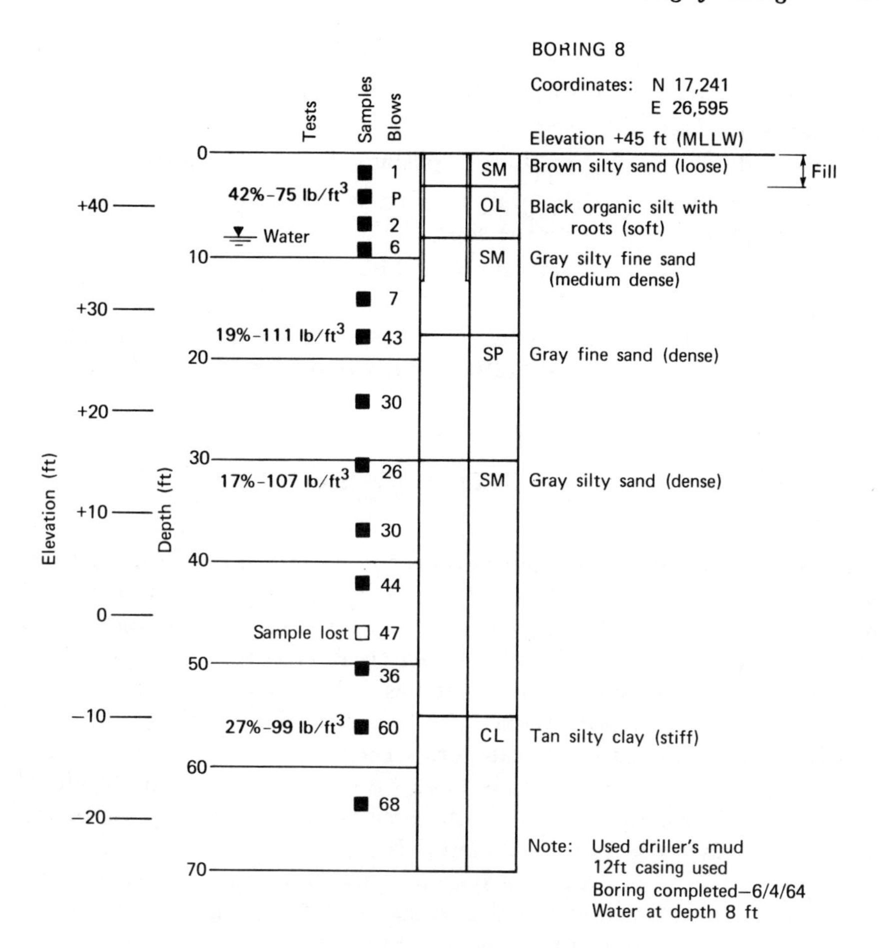

Fig. 13.1 Typical log of borings.

11. The water level, levels at which water seepage occurs, and any perched water.
12. The level at which drilling water (or mud) was lost into the formation.
13. The thickness of the various soil layers encountered.
14. Any topsoil, organic or peaty material, or man-made fill which overlies the site.
15. The colors of each of the soil layers.
16. A classification of each of these soil layers.
17. The relative firmness or compactness of each soil layer.
18. Materials contained in each layer, such as roots, vegetation, and chemicals such as gypsum.

19. Depths at which samples of the soil were obtained. Also, are the samples cuttings, disturbed cores, or undisturbed cores?
20. The type of sampler used.
21. The method of driving, rotating, or pushing the sampler into the ground.
22. The blow count to drive the sampler, or resistance encountered in pushing the sampler into the soil.
23. Any soil samples that were attempted but lost.
24. Results of laboratory tests frequently are also listed on the log of borings. In Fig. 13.1 the laboratory test data listed for the sample at elevation +42 are moisture content (42%) and dry density of the soil (75 lb/ft^3).

13.4 Soil Classification

A number of systems have been devised for classification of soils. Earlier systems were for agricultural uses. Later, a classification system developed by the U.S. Bureau of Public Roads became popular. It described many soils as loam. Next, the Highway Research Board created a system for classifying soils in eight categories listed as A-1 through A-7, *30*.

The Federal Aviation Agency (FAA) prepared a somewhat different classification of soils, setting 11 categories listed as E-1 through E-13, where E-13 is extremely bad soil or muck. The FAA is moving to use the "unified" system, described below, but the FAA classification system may still be encountered in use or in reference materials.

In an effort to improve the system of soil classification, a system was designed and titled "the unified soil classification system." Figure 13.2 presents an abbreviated version of this classification system, *31*.

Basically, the requirements for describing and classifying soils include the following items:

Color.
The grain size of the major portion of the soil.
The grain sizes of other portions of the soil.
The firmness of the soil: whether soft, moderately firm, firm, or stiff; whether the soil is compact or loose.
The moisture content: whether wet, moderately dry, or dry.
Apparent engineering properties.

Figure 13.3 presents a comparison between the unified, Public Roads, and FAA classification systems. For more information, see Ref. *13*.

MAJOR DIVISIONS			GRAPH SYMBOL	LETTER SYMBOL	TYPICAL DESCRIPTIONS
COARSE GRAINED SOILS	GRAVEL AND GRAVELLY SOILS	CLEAN GRAVELS (LITTLE OR NO FINES)		GW	WELL-GRADED GRAVELS, GRAVEL-SAND MIXTURES, LITTLE OR NO FINES
				GP	POORLY-GRADED GRAVELS, GRAVEL-SAND MIXTURES, LITTLE OR NO FINES
	MORE THAN 50% OF COARSE FRACTION RETAINED ON NO. 4 SIEVE	GRAVELS WITH FINES (APPRECIABLE AMOUNT OF FINES)		GM	SILTY GRAVELS, GRAVEL-SAND-SILT MIXTURES
				GC	CLAYEY GRAVELS, GRAVEL-SAND-CLAY MIXTURES
	SAND AND SANDY SOILS	CLEAN SAND (LITTLE OR NO FINES)		SW	WELL-GRADED SANDS, GRAVELLY SANDS, LITTLE OR NO FINES
MORE THAN 50% OF MATERIAL IS LARGER THAN NO. 200 SIEVE SIZE				SP	POORLY-GRADED SANDS, GRAVELLY SANDS, LITTLE OR NO FINES
	MORE THAN 50% OF COARSE FRACTION PASSING NO. 4 SIEVE	SANDS WITH FINES (APPRECIABLE AMOUNT OF FINES)		SM	SILTY SANDS, SAND-SILT MIXTURES
				SC	CLAYEY SANDS, SAND-CLAY MIXTURES
FINE GRAINED SOILS	SILTS AND CLAYS	LIQUID LIMIT LESS THAN 50		ML	INORGANIC SILTS AND VERY FINE SANDS, ROCK FLOUR, SILTY OR CLAYEY FINE SANDS OR CLAYEY SILTS WITH SLIGHT PLASTICITY
				CL	INORGANIC CLAYS OF LOW TO MEDIUM PLASTICITY, GRAVELLY CLAYS, SANDY CLAYS, SILTY CLAYS, LEAN CLAYS
				OL	ORGANIC SILTS AND ORGANIC SILTY CLAYS OF LOW PLASTICITY
MORE THAN 50% OF MATERIAL IS SMALLER THAN NO. 200 SIEVE SIZE	SILTS AND CLAYS	LIQUID LIMIT GREATER THAN 50		MH	INORGANIC SILTS, MICACEOUS OR DIATOMACEOUS FINE SAND OR SILTY SOILS
				CH	INORGANIC CLAYS OF HIGH PLASTICITY, FAT CLAYS
				OH	ORGANIC CLAYS OF MEDIUM TO HIGH PLASTICITY, ORGANIC SILTS
HIGHLY ORGANIC SOILS				PT	PEAT, HUMUS, SWAMP SOILS WITH HIGH ORGANIC CONTENTS

NOTE: DUAL SYMBOLS ARE USED TO INDICATE BORDERLINE SOIL CLASSIFICATIONS.

Fig. 13.2 Unified soil classification system. For more information, see Ref. *3*, Chap. 1, and *31*.

Approximate Equivalent Groups of Unified Soil Classification System, Revised Public Roads System, and Civil Aeronautics Administration System

Unified system	Public roads	Civil Aeronautics Administration	
GW	A–1–a	Gravelly soils not included directly. Note upgrading permitted.	
GP	A–1–a		
GM	A–1–a, A–2–4 or 5		
GC	A–2–6 or 7		
SW	A–1–b	E–1, 2 or 3	E–4 or 5 (usually SM or SC)
SP	A–3		
SM	A–1–b, A–2–4 or 5		
SC	A–2–6 or 7		
ML	A–4	E–6	
CL	A–6, A–7–5		
OL	A–4, A–7–5	E–6	E–8 (usually L group) E–9 (usually not CH)
MH	A–5	E–10, 11 or 12	
CH	A–7		
OH	A–7		
Pt	------------------------------	E–13	

Note: Groups are only approximately equivalent, since different limiting values are used in each system.

Fig. 13.3 Comparison of unified, Public Roads, and FAA classification system.

13.5 Water Level

The water level underlying a site is very important in planning excavations or in constructing basements.

Normally, logs of borings show the water level. This is the water level encountered at the time the borings were drilled.

The water level may be different at the time of construction. Recent rains may cause the groundwater level to rise. The water level may be lowered after a long period of low rainfall. The water level measurements shown on the boring logs may have been done carefully, or they may reflect a single observation during drilling. Sometimes it is necessary to leave a boring open for a period of time, and perhaps insert a perforated casing and then bail down the fluid in a boring to obtain a true measure of the water level. Sometimes it is necessary to flush out a boring several times to remove the drilling mud.

Frequently, water level information can be obtained from other sources. This information may include water levels in nearby water wells and seepage of water into basements in nearby buildings.

13.6 Recommendations

Most reports contain recommendations for the design of foundations. Some comments may also be made regarding construction procedures.

It is important for the contractor to understand the intent of the recommendations. For instance, it may be recommended that foundations be placed at a depth of 5 ft on a gray sandy clay layer. Which is more important, founding at a depth of 5 ft or founding on the sandy clay? If the sandy clay is 7 ft deep over portions of the site, should the foundations be placed at a depth of 7 ft, or may they also be placed at a depth of 5 ft? The report might recommend that slopes cut into the soil be at an angle 1½ horizontal to 1 vertical. This slope, however, might refer to permanent slopes. Temporary slopes might be 1:1, or perhaps ¾ horizontal to 1 vertical. Permanent slopes should have a higher factor of safety, and generally are flatter for erosion and maintenance purposes than may be necessary for a temporary construction slope.

13.7 Construction Planning

It is to the advantage of the contractor to have the soils engineer provide some input during construction planning. Questions regarding slope angles and foundation depths can be clarified. The approach to various phases of construction can be reviewed. Many times the on-site soil after it is excavated can be reused as compacted fill around the perimeter of the basement. The soil may need to be protected from rain. Also, it may be wet when excavated or need to be dried out. In some cases, however, the excavated material may be very difficult to work. In these cases, compaction tests performed by the foundation engineer would indicate a potential problem (see Chapter 12, Section 12.9). In such cases, it may be to the advantage of the contractor to plan on disposing of the excavated material and importing backfill which is easy to compact.

13.8 Limitations and Professional Liability

The soils report generally is an expression of opinions and recommendations by a professional person. The report is not a warranty or an insurance policy.

Recently, more and more lawsuits have been filed against engineers. Soils engineers have been particularly vulnerable. The soils engineer generally is called on to reach a reasonable answer with far less real information regarding the site than he or she would like to have.

Generally, the ground for a reasonable suit is carelessness. It is recognized that opinions and recommendations are of a variable nature—not subject to precise calculations. Therefore, some recommendations are based on judgment. Judgment may be different between various professionals, and who is to say that the judgment of one engineer is in error?

The matter of professional liability needs consideration in much more detail than is possible in this book.

14

Field Tests

14.1 Bearing Tests

The strength of soil at a proposed foundation level is sometimes tested by making static load tests on model footings. A pit is excavated to the depth proposed for testing. The pit generally is made considerably wider than the area of the test footing, usually four to five times the width of the test footing. Test footings generally are at least 1 by 1 ft in size, and so the pit would be at least 4 by 4 ft in size.

A loading platform or reaction frame can be constructed over the plate, or for convenience sometimes a loaded truck trailer is positioned over the test footing. A jack is used to apply load on the test footing. Dial gauges or a surveyor's level is used to measure settlements of the footing as load is applied.

Usually, load is applied in several steps. A plot of load versus settlement is made for each increment of load applied. The test is continued until settlement becomes progressively larger under each increment of load, or until the capacity of the testing apparatus is reached. Readings of rebound are taken as the load is removed. A typical test setup is shown in Fig. 14.1. A typical load-settlement curve is also shown in Fig. 14.1. A detailed specification for such tests is contained in ASTM, test designation D1194, *32*.

Usually two or more load tests are performed. Two sizes of test foundations may be used, such as 12 by 12 in. and 17 by 17 in. (2 ft^2).

The results of the tests can be examined to determine a point at which yield of the soil starts. The indicated bearing pressure usually is divided by two, three, or more (to provide a factor of safety) to determine a "design bearing value."

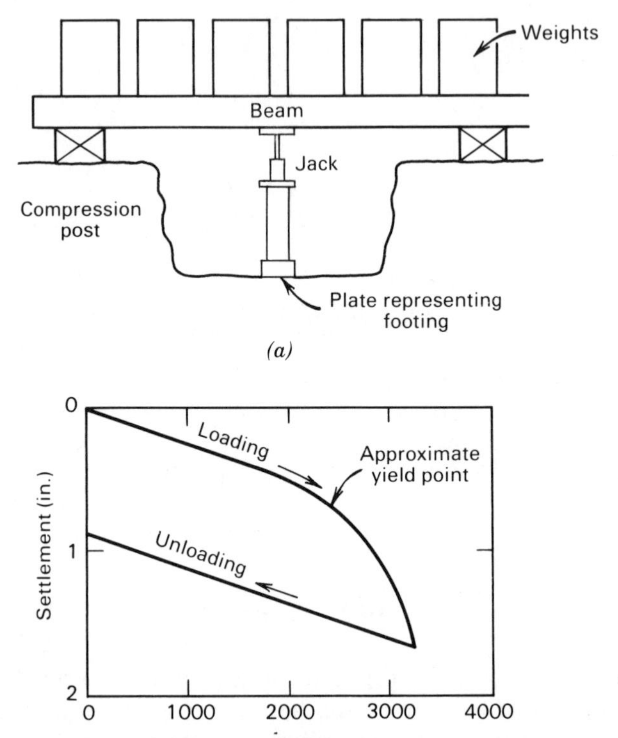

Fig. 14.1 (*a*) Soil bearing test. Settlement of the footing is read with a surveyor's level or with dial gauges mounted on an independent reference beam. (*b*) Typical graph of load vs. settlement and rebound derived from test.

Limitations to this method of test are as follows:

1. The soil tested extends to a depth of one to two times the size of the test footing. A true full-size footing would stress the soils to greater depth. A soft layer at a depth of 4 ft could affect a large footing. Therefore, borings or some deeper information is also necessary to go along with load test data.
2. The test results indicate primarily bearing capacity, and are not a reliable indication of likely settlement of proposed foundations. As an example, if a test footing 2 by 2 ft in size settled 1 in. at the proposed design bearing pressure, a real footing 8 by 8 ft in size would be expected to settle about 4 in. A footing 4 by 4 ft in size would be expected to settle approximately 2 in.
3. Settlements of clay soils occur slowly, over a long period of time. Long-term settlements of building foundations may be much larger than would be estimated from a load test.
4. A change in water level could affect the bearing capacity of the soil.

Soil bearing tests are made on the subgrade soils for airport runways and highways. Static load tests are made on circular plates. The test plate usually is 30 in. in diameter. Specific test procedures have been developed and are described in ASTM D1196, *33*. Interpretation of test results and use of the data are much different from footing load tests previously described.

14.2 Vane Shear Tests

Vane shear tests can be made in test borings. Usually they are performed in clay soils, particularly soft clays free of sand layers or gravel.

The test boring is stopped at the depth proposed for testing and is cleaned out. The vane apparatus is lowered to the bottom of the borings, and pressed into the soil below the bottom of the boring. A sketch of a vane test apparatus is shown in Fig. 14.2.

The vane is rotated at a constant rate, and the force required to cause rotation is measured until the strength of the soil is exceeded.

By using the dimensions of the vane and the force required to cause rotation, the stress required to shear the soil is determined.

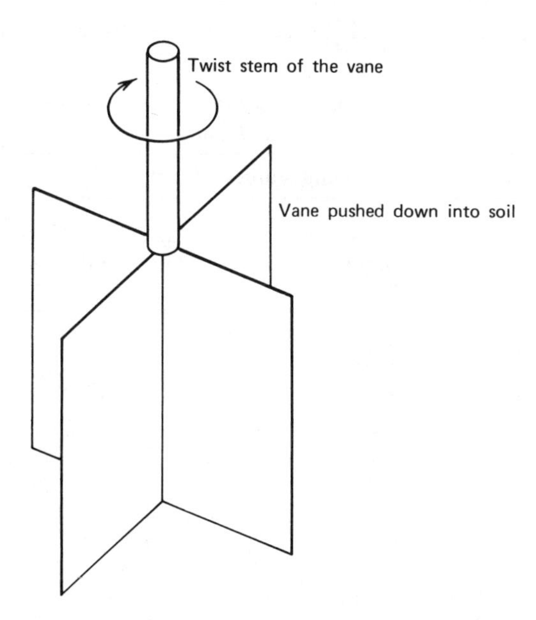

Fig. 14.2 Vane test device. For example, if the vane is 1 in. in diameter and 1 in. high, then shear resistance in lb/in.^{2} $\sim$ ½ torque in in.-lb to turn the vane. Therefore, if torque is 2 lb at the end of a 10-in. wrench, the torque is 20 in.-lb and the cohesion is 10 lb/in.^{2}. This represents a fairly strong soil.

14.3 Penetrometer Tests

A variety of tests have been devised to measure the resistance of the soil to penetration. This might include a penetration of

The boring casing.
A sample pipe.
The 2-in. diameter "standard penetrometer."
A small-diameter cone.
Soil samplers of various sizes.

Penetration of these devices in the soil may be caused by pushing on the drill pipe or rods with hydraulic jacks, or by hammering on the drill rods using a falling weight as a driving hammer.

These results give a qualitative indication of the soil characteristics and strengths. Usually, these strength indications are used to correlate with laboratory test data. However, in some cases the results of penetrometer tests are used directly to estimate soil bearing capacity or the probable penetration of piles into the soil.

There is no generally accepted conversion from blow counts (N values) to bearing values for a soil, and there is considerable difference in opinion as to whether it is even practical to attempt to use blow counts for this purpose. However, Table 14.1 is one suggested conversion table. It may be

Table 14.1 The Standard Penetration Test (N) *as an Indication of Bearing Capacity of Soils*

	Bearing Values (lb/ft²)		
		Sandy Soils	
N	Clay and Silty Clay Soils	Fine Sands	Coarse Sands
2 blows[a]	500	—	—
4	1,200	—	—
5	1,500	500	1,500
6	2,500	700	1,800
8	4,500	1,200	2,000
10	5,500	1,500	2,500
12	6,000	1,800	3,000
15	7,000	2,000	3,500
20	8,000	3,000	4,500
25	9,000	4,000	6,000
30	10,000	4,500	6,500
40	12,000	5,500	8,000
50	14,000	6,000	9,000
60	15,500	6,500	10,500
70	17,000	7,000	12,000

[a] Only to be used for unimportant structures with light loadings.

of some guidance, but *should not be used alone* as the basis for selecting a design bearing value.

Blow counts certainly can help a contractor estimate the difficulty in making excavations and in driving piles and sheet piles.

These values do not consider submerged conditions, deep foundations, or mat foundations. These values are considered "safe" bearing values, rather than almost at the failure point.

14.4 Menard Pressure Meter

This device is a special type of balloon operated in a way that is somewhat similar to the way a hydraulic jack is. It is lowered down the boring to the desired depth of testing. The diameter of the pressure meter is only slightly smaller than the borings. Guard cells on either side of the balloon are first inflated so that they press against the walls of the boring. Then the balloon itself is expanded by hydraulic pressure. Hydraulic pressure is applied inside the balloon until it forces itself out sufficiently to cause failure of the soil surrounding the boring.

Special techniques are required for interpretation of the test results.

14.5 Air-Operated Percussion Drills

Occasionally, percussion drills are used as a crude method of probing firm soil or rock. The rate of penetration is some indication of the firmness of the material. Fractures, voids, and other soft conditions can be discovered.

14.6 Pocket Penetrometer

A small device which is frequently used to make approximate measurements of the strength of clay or silty soil is called the pocket penetrometer. It can be used on the job site to measure the strength of soil in trenches, of soil samples obtained from borings, or of soil on the bottom of footing excavations (see Ref. *34*).

14.7 Pile Load Tests

The types of piles proposed for use are frequently driven at a site for testing. The testing may be to verify "design pile capacity" or to determine pile lengths for precasting of concrete piles. Pile load tests are run in a manner fairly similar to that for bearing tests on soil. A typical test setup and load settlement diagram are shown in Fig. 14.4. Specifications for such tests are

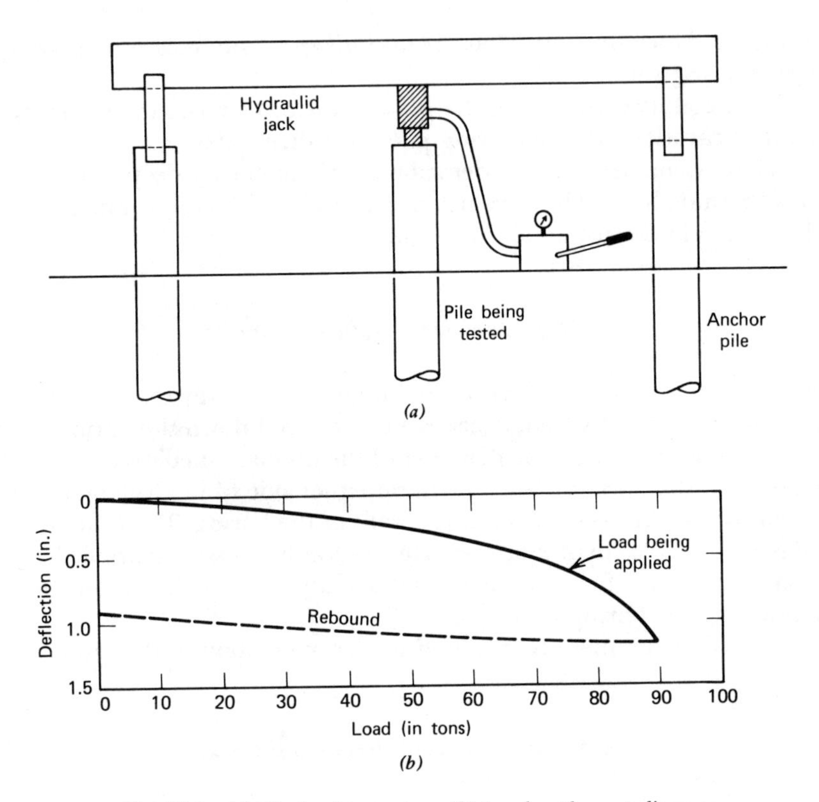

Fig. 14.3 (*a*) Pile load test setup. (*b*) Load settlement diagram.

given in Ref. *35*. Also, most building codes specify procedures for running the tests and evaluating the results (see Ref. *36* and Ref. *37*, p. 240).

In evaluating test results, bear in mind that (*a*) a group of several piles performs differently than a single pile. Settlements are usually more. A bearing stratum may be underlaid by a soft layer. This may not show up in testing one pile, but could result in settlement of a group of piles. (*b*) Long-term settlements may be more than settlements measured during the test. (*c*) Production piles must extend into the same bearing stratum (for end-bearing piles), even though the bearing stratum may be deeper (or shallower) in other parts of the site. For more discussion, see Chapter 20, Section 20.4.

14.8 Dry Density

Frequently, the density of soil is measured in the field. This can be done by taking samples of the soil by pushing or driving in a soil sampler. This procedure is commonly used for testing compacted fills and is described in Chapter 12, Section 12.3.

In testing fills, however, it is more common to dig a hole into the compacted soil to obtain a sample of the soil. The volume of the hole must be measured. An alternative is nuclear testing. These methods are described in Chapter 22, Section 22.13.

Frequently, moisture content of the soil is also tested in the field. The method is described in Chapter 12, Section 12.2.

14.9 Electrical Resistance

Soils resist the flow of electricity through them. Usually, dry, firm soils have a high resistance. Wet and soft soils have a low resistance. Also, soils containing chemicals have a low resistance. Resistance is important and must be taken into account when considering such problems as ground of electrical transmission lines and the need for protection of foundations and buried structures from "hot" soils.

Electrical resistance is frequently measured by a device which pumps electrical energy into the ground through a steel stake and measures the flow to another steel stake. A device for making such measurements is shown in Fig. 14.4.

This kind of survey, taking several measurements, can be used to estimate the depth to bedrock (see Ref. *18*, p. 445).

14.10 Water Level Measurements

It may be necessary to measure groundwater levels for months or years. Commonly used methods are an open dug pit or drilled hole, installation of cased wells, and installation of piezometers.

In pits or wells a tape or electric probe device can be lowered to measure the depth to freestanding water.

The piezometer is a pressure device which measures water pressure in any selected layer of soil. Therefore, a piezometer can measure any artesian water pressures in a deeper layer of soil. Also, if a clay layer is overloaded by an embankment of new fill, the excess water pressures can be measured. These measurements can warn of likely failures or landslides. A piezometer is shown in Fig. 14.5.

14.11 Measurement of Earth Deformations

Lateral ground movements are more difficult to measure than vertical movements, which can be measured relatively well with surveyors' levels

Fig. 14.4 Device for measuring electric resistivity of soils.

Fig. 14.5 Piezometer (atmospheric type).

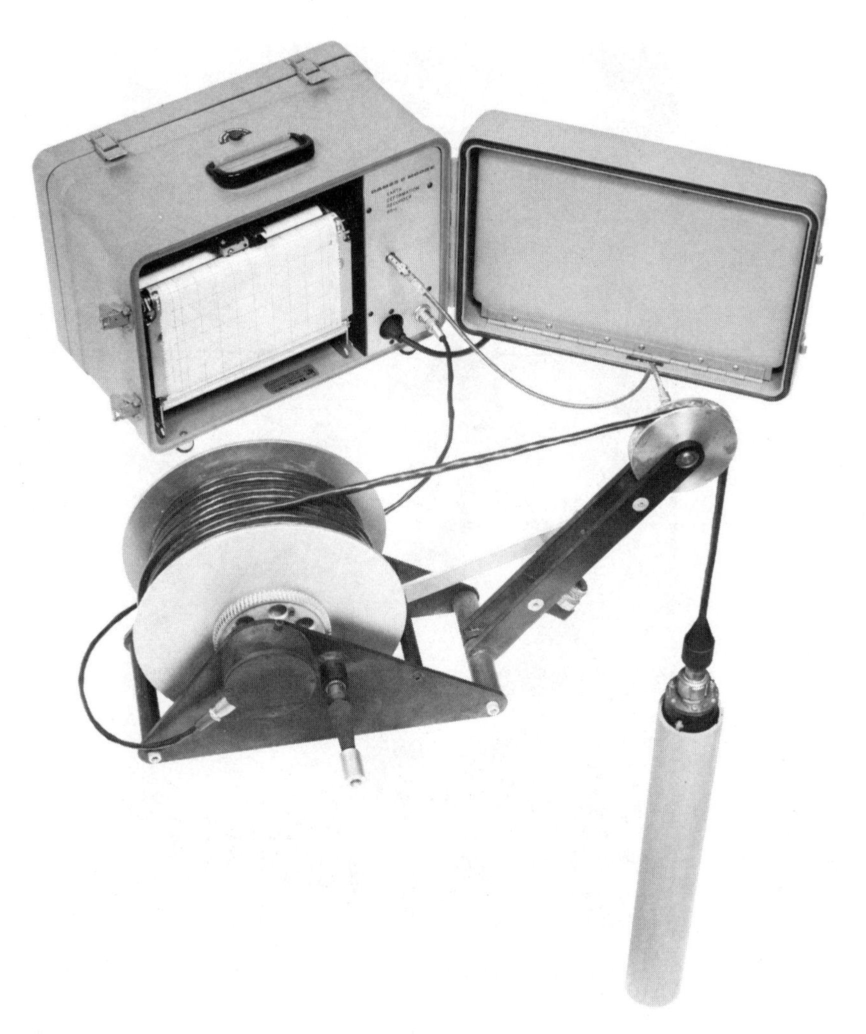

Fig. 14.6 Instrument for measuring earth deformation entering casing.

and other leveling devices. Installed special cased holes are commonly used to measure horizontal movements. Instruments are run down and back up the holes to measure any bending or leaning of the casing. If the casing crosses a landslide slip plane where movement is occurring slowly, the casing bends in an S shape, and this is detected by the instrument when it is lowered down the casing. The most commonly used instrument is the slope indicator (see Ref. *38*, p. 1042). Another typical instrument is the "earth deformation recorder." In Fig. 14.6, this instrument is shown entering a casing. Wheels on the side of the instrument track into grooves which are machined into the casing to orient the instrument.

15

Excavations

Most general contractors do well on vertical construction above grade. However, many have difficulties underground and may therefore lose money on their project.

Frequently, the project manager is selected for his or her ability to erect steel, pour concrete, or perform other vertical construction. Too many times the manager does not understand underground problems and gets into serious difficulty before recognizing that assistance is needed.

Earth construction is as difficult and demanding as aboveground construction.

15.1 Slope Stability

15.1.1 Slope Angle

Many excavations are started with a vertical cut. Some soils will stand to considerable depths when cut vertically, although most will not. When vertical slopes slough off to a stable angle, large blocks of material may slide down into the excavation.

In sandy soils, the sand will generally tend to slough and cave in during the process of excavation. Usually, this results in a fairly stable angle, without serious hazard.

However, cemented sands and silty or clayey soils sometimes are excavated to considerable depths before a large block of soil slides into the excavation.

The angle at which soil can be expected to stand temporarily during excavation can be calculated. Some rough rule-of-thumb slope angles are presented in Table 15.1.

In starting an excavation, it is easiest to start the cut at the proposed slope angle. If the desired temporary slope angle is one horizontal to one vertical and the depth of excavation is to be 15 ft, the top of excavation should be started 15 ft outside of the proposed toe of excavation.

15.1.2 Undermining

After a slope has been cut, it sometimes is necessary to undermine the slope to get in certain foundations or utilities.

Where undermining is necessary and it is not possible to lay back the slope at a stable angle, undermining should be done in narrow sections or windows. The top of the slope should be unloaded as much as possible, and construction equipment should not be on the top of the slope.

Sometimes, slopes are undermined in the process of excavation. The soil caves and runs to the excavation machine. This is not too hazardous in clean sand, which will slide back to its angle of repose, but this can be hazardous in cemented soils, damp sands with apparent cohesion, or silts or clays, which will stand vertical temporarily, and then a large block will break off and come down like a landslide.

Table 15.1 Some Typical Temporary Slope Angles[a]

Soil Type	Temporary Slope Angles
1. Sand or sand and gravel	45° for damp slopes; 35° or 1½ to 1 for dry slopes; flatter for wet slopes.
2. Cemented sand	Vertical to 10 ft; ½ to 1 to 20 ft or more; ¾ to 1 for high slopes.
3. Soft silt or soft clay	Vertical to 3 ft; ½ to 1 to 6 ft; ¾ to 1 to 10 ft; 1½ to 1 for high slopes. Flatter slopes, like 4 to 1, for wet slopes.
4. Moderately firm silt or clay	Vertical to 6 ft; ½ to 1 to 10 ft; ¾ to 1 to 20 ft; 1 to 1 for higher slopes, except flatter slopes for wet soil conditions.
5. Firm silt or clay	Vertical to 10 ft; ½ to 1 to 20 ft; ¾ to 1 to 30 ft; 1 to 1 for higher slopes, except flatter slopes for wet soil conditions.
6. Mud	Silt or clay soils containing high amounts of moisture may require very flat slopes, like 4 to 1 or 6 to 1 or flatter, unless dewatering or drainage is used to reduce the water problem.

[a] There are many exceptions—the "typical values" above are not intended for use for design of slopes. The Federal Act entitled "Occupational Safety and Health Act," as well as state industrial codes, limits the height of unbraced vertical cuts where workers are located. Typical slope angles for permanent slopes are given in Chapter 21, Section 21.2.

15.1.3 Erosion

Excavation slopes have been denuded of natural cover and protection, and are usually very susceptible to erosion in heavy rains.

It is particularly important to limit the amount of water running down over the excavation slope. This can be done by building a curb or dike at the top of the slope, which forces water to run away from the excavation rather than over the excavation slope.

The face of the excavation also can be protected prior to an imminent rainfall by covering with plastic sheets or by spraying with various waterproofing materials. Sodium silicate is one material used for this purpose (a modification of the Joosten process). This material can be injected into a slope or sprayed on the surface (see Chapter 28, Section 28.5).

15.1.4 Cracking

Cracking of steep excavated slopes may become a problem, as a result of severe drying of the soil. The soil tends to ravel and run down the slope. Even worse, deep cracks may allow blocks of material to fall out of the slope. Often slopes appear to be at a stable slope, but a block of material can drop off the wall of the excavation. Much damage and injury result from this one problem.

Later, when backfilling is completed, the soil may become moist again. The soil may swell and possibly crack concrete walls or concrete slabs placed over the slope.

Cracking can most easily be reduced by limiting the evaporation of water. Occasionally this is done by frequent sprinkling or "fog spraying" of the slope. This also can be accomplished by placing a protective cover over the slope. Protective covers may consist of plastic sheets, sprayed-on chemicals, sprayed-on bitumastic materials, or other waterproofing materials (see Ref. *39*).

15.1.5 Loads on Top of Slope

It is natural for the contractor to put machinery, excess excavated soil, or construction materials at the top of the excavated slope. In addition, heavy machinery may produce vibrations, increasing the hazard to the slope.

The stability of a slope with an added surcharge load at the edge can be analyzed, as described in Chapter 24, Section 24.2. Also, building codes, or the federal and state safety codes, may spell out limitations to loading the edge of an excavation. Also, it is possible to calculate a "safe distance back" of the edge for placing loads.

As a crude guide, the safe distance back from the top of a slope can be estimated by assuming that the slope could be considered safe if it were higher but placed at the slope angle used in making the excavation. If this slope angle is ¾ to 1, for instance, and the weight on the tracks of a piece of equipment is 500 lb/ft^2, this would be equivalent to making the excavation 5 ft deeper. This is shown in Fig. 15.1(*a*). Since the equipment may be used for making lifts, the maximum pressure on the tread closest to the excavation may be increased. This increase should be added to the track load in making the calculation. If, for instance, the track load is increased to 1000 lb/ft^2 during a lift, the equivalent height of new soil is 1000 lb/ft^2. The 10 ft of soil can be replaced with a block of soil, as shown in Fig. 15.1(*b*). The edge

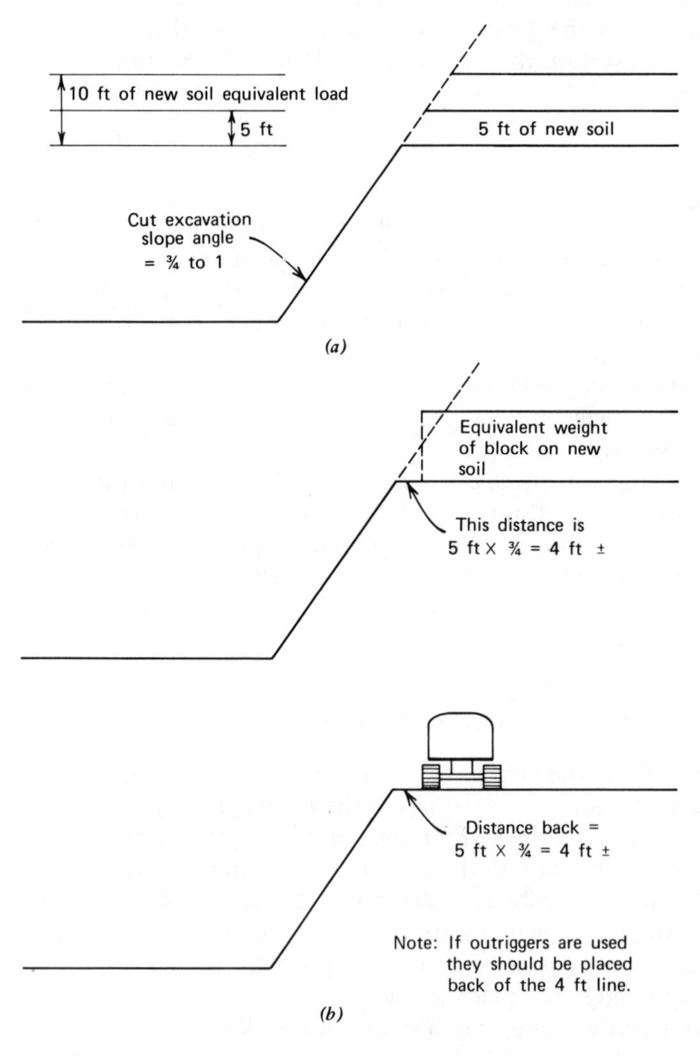

Fig. 15.1 Guide to safe distance back from top of slope for storage of materials or placing equipment.

of the block is approximately 4 ft back from the edge of the slope. The crane track also should be kept at least 4 ft back from the edge of the excavation.

15.1.6 Detection of Movement

Landslides and slope failure rarely occur without some signs of impending failure. The primary problem is that in many cases these indications are not observed or heeded.

During operation of equipment in a precarious position at the top of a slope, someone should be given the specific assignment of frequent inspection of the top and the slope. This person should look for signs of cracking back of the top of the slope back to a distance that is equal to the height of the slope. Also, he or she should watch for bulging at the center or toe of the slope and for soil particles running off the slope below the piece of machinery.

Small slope movements can be detected by surveying on fixed reference points. A more convenient but more expensive method is to install special casings in drilled holes. These casings and instruments to measure movement of the casings can be obtained from several sources. However, it would be best to work with a soils engineer familiar with the operation of this equipment (see Chapter 14, Section 14.11).

15.1.7 Vertical Cuts

In some cases, vertical cuts can be made in soils that are cemented or are stiff, cohesive silts or clays. In addition, vertical cuts sometimes are made in sandy soils exhibiting temporary "apparent cohesion" due to water in the sand.

Vertical cuts should nearly always be considered temporary and should be backfilled or otherwise stabilized in a short period of time.

In general, the maximum vertical height to which a silt or clay soil can stand is equal to the following expression:

$$\text{Height} = \frac{2 \times \text{cohesion}}{\text{soil weight}}$$

Cohesion is measured in laboratory tests, as described in Chapter 12, Section 12.5, and should be described in the soils report.

For a clay having a cohesion of 500 lb/ft^2 and a weight of 100 lb/ft^3, the temporary height at which a vertical bank could be cut is 10 ft. This calculation includes no factor of safety. With a safety factor of 1.5 the allowable height of cut is 6½ ft.

In general, vertical cuts should be avoided. Equipment operators should not start an excavation digging vertically, and then after the excavation is completed find that it would be extremely difficult to go back and flatten the slope. Excavations should be started by digging out to the required cut lines for the desired slope.

15.1.8 Slope Stabilization

If a slope is cut too steep and starts to fail, the obvious method of stabilizing is flattening the slope. If there is not room to flatten, some other action is necessary. Hopefully, this action will be taken before excavation has gone too far.

Stabilization methods include

Dewatering. In many cases, dewatering substantially back from the slopes using well points or wells increases the stability of the slope.

Guniting. If the slope is not badly oversteepened, a gunite surface will sometimes hold in soil moisture and add some strength.

Chemicals. Where steep slopes must be cut in areas of tight space limitations, it is possible to stabilize certain soils by injecting chemicals. See Chapter 28, Section 28.5.

Bracing. Slopes are most often stabilized by shoring and bracing, as described in Chapter 16.

15.2 Bottom Protection

15.2.1 Drying and Cracking

Soils at the bottom of an excavation may become excessively dry and crack and shrink if they are exposed for a long period of time to dry, hot weather. This may create a problem after the building floor slabs are poured if the soils swell when they regain their natural moisture content. Therefore, the soil may need to be sprinkled or temporarily covered with sand or plastic sheets to limit evaporation.

15.2.2 Saturation

More frequently, the soils at the bottom of an excavation may be saturated. Even though the groundwater level may be below the bottom of the excavation or the site may be dewatered by rim trenches, wellpoints, or wells, it is common to find that the bottom of an excavation becomes unstable for operation of front-end loaders, dump trucks, and other heavy-wheeled equipment.

Also, soils which are saturated and soft may be undesirable soils on which to pour spread foundations for support of the building columns. Such soils may have swelled or rebounded on removal of the overburden soils. The soils therefore would recompress on application of the foundation load. In such cases where the subgrade is in wet clay or silty soils, it frequently is less expensive to overexcavate 12 to 18 in. and backfill with select free-draining soils to provide a "working course." Such backfill must be well compacted.

Occasionally, equipment breakdown may cause the dewatering operation to stop for hours or even days. In this case, the groundwater rises upward and may fill the excavation. During the time the water is flowing upward into the excavation, the soils on the bottom of the excavation may be severely loosened and disturbed.

When the dewatering system is again in operation and the excavation has been dewatered and dried out, it may be necessary to remove the loosened soil. The excavations may be backfilled with select and well-compacted soil. If the existing soils at the bottom of the excavation are sandy soils, similar to that which might be imported as "select soil," it is possible that the loose soils can be recompacted adequately. The entire bottom of the excavation should be recompacted.

A common problem in the case of sands with few fines, whether in-situ or imported, is "bulking" of the sand. At commonly occurring, neither dry nor saturated soil moistures, the capillary moisture causes the sand grains to adhere in apparent cohesion and bulk. As a result the material rather strongly resists compaction. This too frequently leads to low density, which later increases on saturation, with consequent loss of support of foundations or slabs.

Heavy rainfall or freezing also can soften the bottom of the excavation. Rainfall protection should include curbs or small dikes around the perimeter of the excavation to prevent the job site from becoming a sump for the whole neighborhood. An extreme treatment by one contractor involved erection of a circus tent over the site during the rainy season. Freezing is more difficult to protect against. Sandy soils are not affected much, but silty or clayey soils can expand when they freeze and turn into mud when they thaw. It is very bad to build foundations on frozen ground, for they settle a lot the next spring when it thaws. If it is not possible to prevent freezing by covering with soil or by heating, then it may be necessary to excavate all frozen soil below footings and replace it with gravel or lean concrete. These costs should be figured into the bid. Note also that excavated frozen soil is virtually impossible to replace satisfactorily as embankment.

15.2.3 Heaving

During driving of pile foundations, the excavations for pile caps have in many cases been found to heave upward. Figure 15.2 shows the excavation for the foundation for a tall stack. Approximately 70 piles were driven at a spacing of 3½ ft center-to-center. The piles were approximately 60 ft long, step tapered, with an average diameter of 12 in.

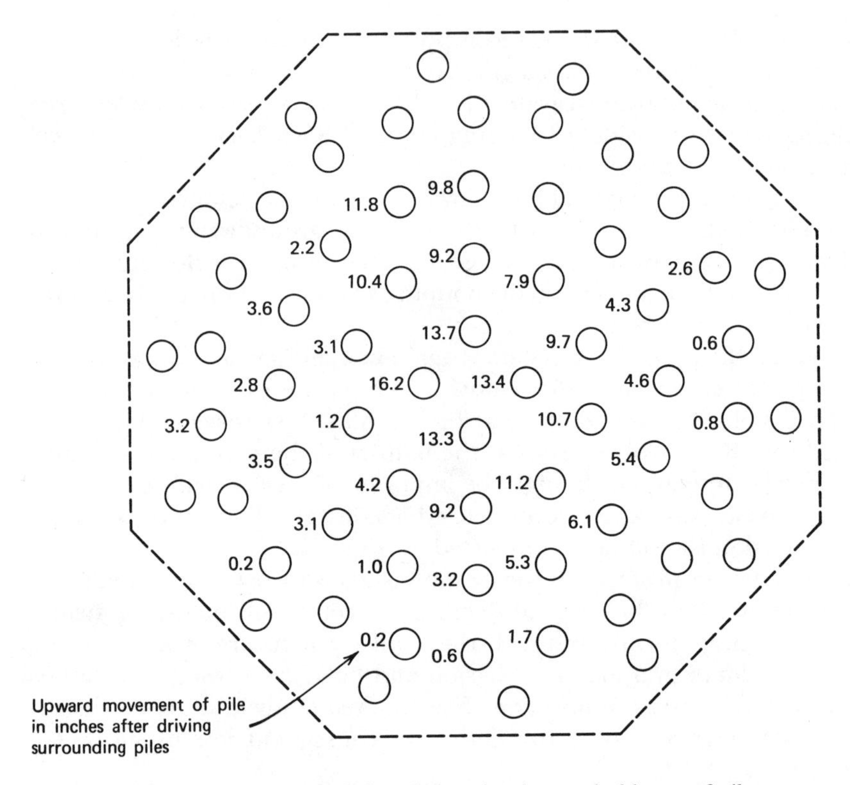

Fig. 15.2 Smoke stack foundation showing vertical heave of piles.

During driving, the ground was observed to heave upward. The ground also lifted up previously driven piles. By the completion of driving, uplifting of piles was measured and varied from 0 to 16 in. Average heaving was around 4 to 5 in.

In many cases, the pile cap is overexcavated to accommodate ground heave. Also, predrilling may be used at each pile location to remove excess soil and thus prevent heaving. Heaving is described in more detail in Chapter 20, Sections 20.8, 20.9, and 20.10.

15.2.4 Accidental Overexcavation

Occasionally, an excavation is made too deep, either by surveying error or by mistake on the part of equipment operators. Also, excavations may be cut deeper so that the corners can be cut out with large machinery.

It is a natural tendency for equipment operators to backfill overexcavated spots by dragging bulldozer blades or loader buckets across the bottom of the excavation, thus disguising areas which have been overexcavated. This cannot be allowed, as discussed further in Chapter 18, Section 18.10.

15.3 Selection of Equipment

The selection of equipment for excavation is important to the economics of the work. Soil characteristics affecting the choice of equipment include

Hardness.

Stickiness or cohesiveness.

Amount of water present.

Ability to support heavy equipment without its rutting and becoming mired.

Some indication of the hardness of soils to be encountered in an excavation can be obtained from the soil investigation report. Indicators are as follows:

The description of the soil.

The sampler blow count.

The plastic nature of the soil.

The design bearing value for the soil.

The recommended slope angle for excavations.

The method of excavation may be important in some cases. The soil generally lies in layers. It may be desirable to excavate topsoil off as one procedure and secondly to excavate and save the good material, which can be used as a select fill or backfill. By contrast, if a general filling operation is to be done with the material excavated, it may be desirable to cut across layers, thereby mixing the poor and good soils together.

More detailed discussions of selection of equipment are given in Ref. *40*.

15.4 Dewatering

15.4.1 Symptoms of Trouble

As an excavation reaches the groundwater level, it may not be immediately evident that groundwater is being reached. However, bulldozers or wheeled construction or hauling equipment in the bottom of the excavation may develop trouble with rutting or weaving and pumping of the soil. In some cases, it appears that the construction equipment may be in danger of sinking out of sight if it continues to excavate. Occasionally, on removal of equipment from the excavation, water will break through the surface, bringing up soil. The water appears to "boil" as it comes out of the ground, forming small cones which look like volcanoes. Such a "sand boil" is shown in Fig. 15.3.

Fig. 15.3 Sand boil.

15.4.2 Methods

The most common method of dewatering an excavation is to construct a sump or several sumps in the bottom of the excavation. Frequently, sumps are placed outside the proposed building lines. It may be necessary to dig ditches (rim trenches) around the perimeter of the excavation leading to the sumps. As water flows into the sumps, they are dewatered by pumps. Pumps used should be designed for this purpose. Flyght submersible electric pumps are frequently used. Detailed discussion of methods of dewatering is contained in Refs. *41* and *42*. Other methods include well points, wells, gravel blankets, and French drains.

15.4.3 Underground Tanks

Large tanks are placed underground for storage of oil, petroleum products, and other fluids. Frequently, such tanks are placed in an area of high water, requiring dewatering. In such cases, dewatering must be continued so that backfilling of the excavation can be done in the dry.

There have been a number of cases in which tanks have popped out of the ground during backfilling. In many of these cases, water jets were used as a means of compacting the backfill soil. In other cases, water was drawn down by pumping during placing of the tank, and then allowed to run into the excavation during the period of backfilling. If the water level

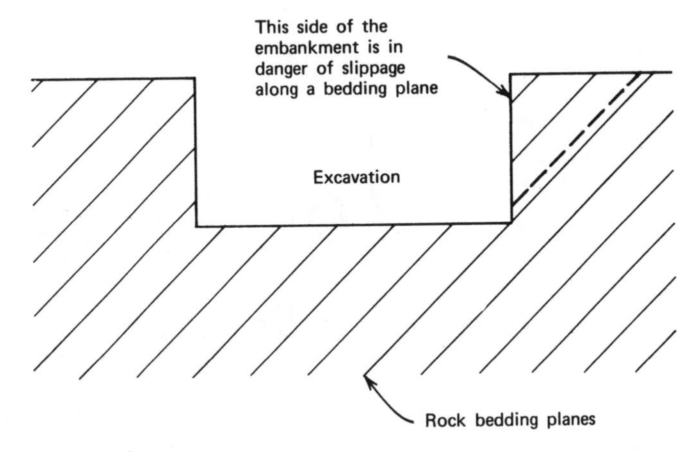

Fig. 15.4 Example of potential slippage of rock along bedding planes.

comes back to its original level, it exerts a considerable upward or buoyance effect on the tank. If the backfill has not yet been completed, sufficient fill over the tanks or tiedowns will be required to prevent the tanks from rising out of the ground when they are emptied.

15.4.4 Underground Pipelines

Excavations for pipelines are discussed in Chapter 17, Section 17.5.

15.5 Rock Excavation

Excavations into rock are not extensively covered in this book.

Many excavations intercept weak planes in rock. These planes may be the bedding planes in sedimentary rock, fractures in hard rock, or fault zones in any kind of rock. If the weak plane lies at an angle, it can cause one of the sides of the excavation trouble, as shown in Fig. 15.4.

16

Shoring and Bracing

The design of shoring and bracing usually is done by the contractor or by a subcontractor specializing in such work. The design generally is the responsibility of the contractor; however, the design is done by the contractor's own engineers or by a specialty subcontractor. In this case, the specialty contractor hires or retains registered professional engineers to prepare the design. The laws concerning responsibility for design vary among states.

Occasionally, the owner may have the shoring system designed and then he or she takes the responsibility for proper performance.

Some building departments have made efforts to set standards for shoring systems and for approval or disapproval of designs proposed by the contractor, because they do not take any responsibility.

OSHA, in some states the Department of Industrial Safety, or the local building department may require approval of proposed shoring systems by their safety engineers. Any of these agencies can close down a job.

The design of shoring systems is based on the expected pressure of a wedge of soil on the back of the shoring. A typical assumed wedge is shown in Fig. 16.1.

16.1 Sheet Piles

Sheet piles may be constructed of steel, reinforced concrete, or wood. Steel sheet piles have interlocks which tie the sheet piles together. Concrete or wooden sheet piles usually have tongue-and-groove connections at their edges, which partially tie the sheet piles together.

Sheet piles are driven at the edge of a proposed excavation; usually the sheet piles are installed before excavation has started, when excavation has proceeded to a moderate depth, or when water is encountered. Sheet piles usually are driven with hammers similar to pile driving hammers, except smaller and usually double acting. Smaller sheet piling may be driven with adaptations to jack hammers.

Usually a template or guide is set on the ground to assist in alignment of the sheets. The permanent wale may be set as the front face guide for the sheets, as shown in Fig. 16.2. Usually the corner piles are set in position first. The other sheets are pitched from each end so that closure is made at the center. Sheets should be driven in lifts, not over 5 ft each, to prevent them from being driven out of their interlocks. In harder ground, the lifts should be shorter.

After the sheet piles are driven and an excavation made down to a depth approximately one-quarter of the proposed excavation, the sheet piles are restrained. The restraint may be developed by cross bracing, rakers, or tieback anchors. Examples are indicated in Fig. 16.2.

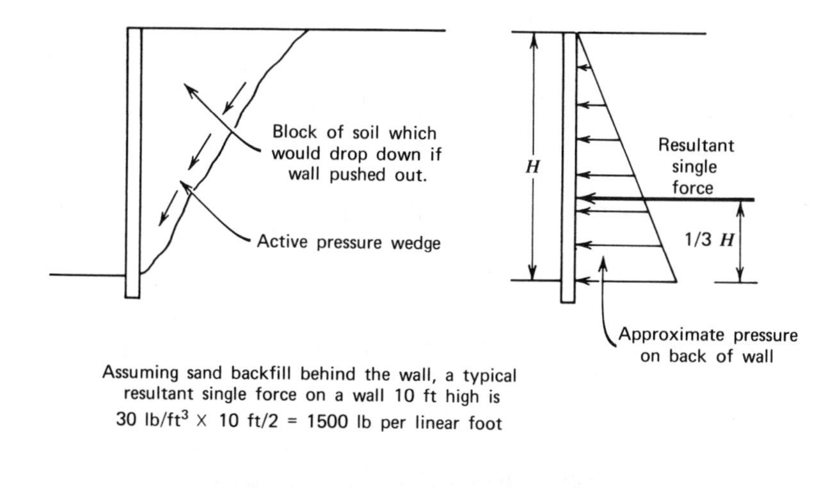

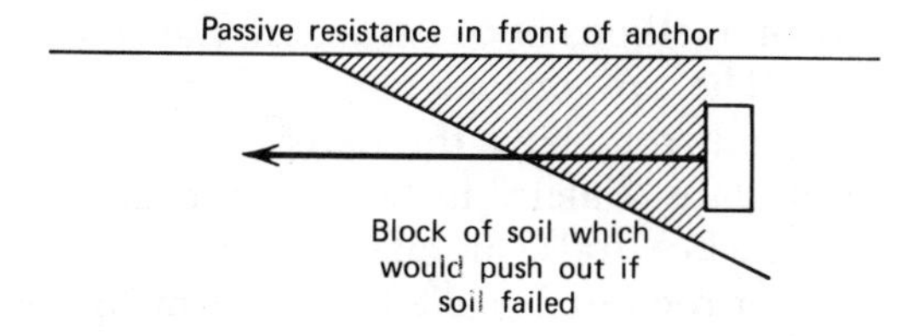

Fig. 16.1 Active soil pressure on wall can be resisted by restraining tieback anchor.

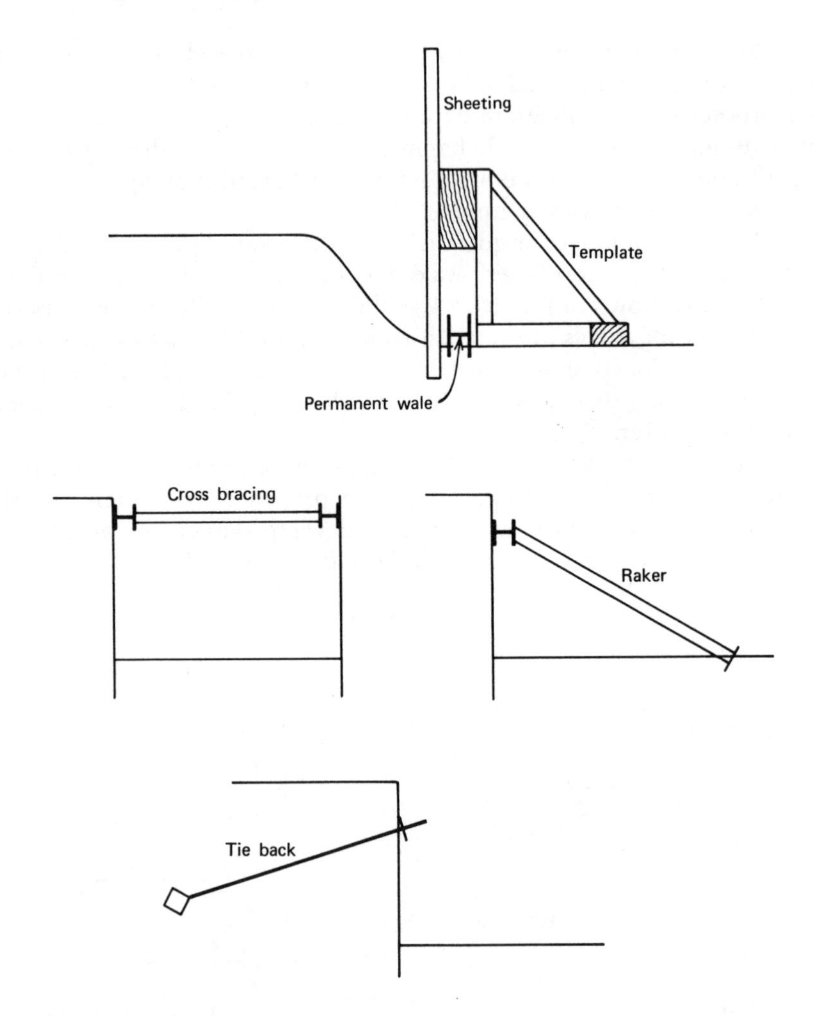

Fig. 16.2 Template and permanent wale used to align sheet piles during driving.

Sheet piling normally is designed by a design engineer. The proportions depend considerably on methods of construction, the type of soil involved, and water conditions. The proportions indicated in Fig. 16.3 are average for reasonably good sandy soil above the water table.

It is common to place "walers" in front of the sheet piles as part of the bracing system. Anchor rods, or internal bracing, push against the walers. When the anchor rods are installed, they normally are pretensioned by turnbuckles on the anchor rods or by other methods. The pretensioning should be approximately equal to the expected force to be carried by the tie rods. Therefore, as excavation proceeds, it would be expected that the

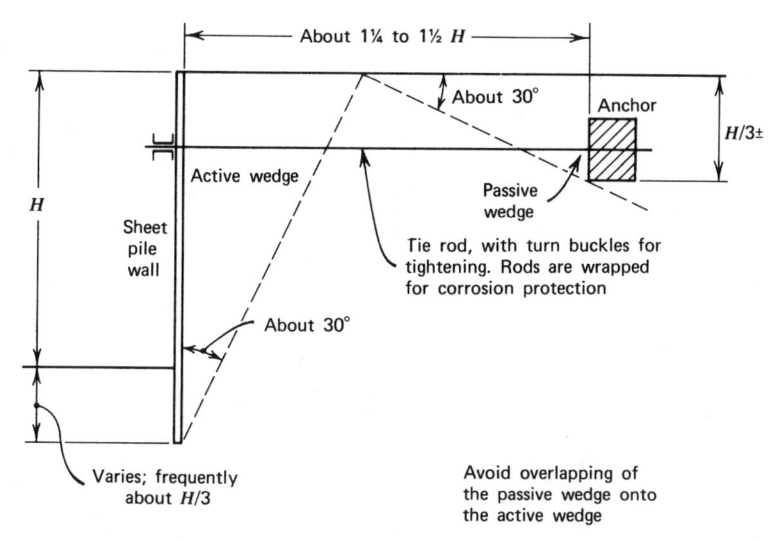

Fig. 16.3 Sheet pile wall with tiebacks and anchors.

deflection of the sheet piles would be limited to bending of the piles, and not to stretching of the anchor rods. This tends to limit lateral deflection and to protect streets or buildings adjacent to the excavation. Some deflection of sheeting is inevitable; therefore, this system is never as positive as underpinning an adjacent structure. If bracing is done by internal inclined posts, sometimes called "rakers," or "kickers," it is common to use jacks to preload the rakers so that they push against the walers with a force equal to the expected design force. In this manner, deflections are limited as the excavation proceeds.

16.2 Soldier Piles

Soldier piles frequently are more economical than sheet piles and are commonly used in excavations for buildings.

Soldier piles are strong beams placed vertically along the perimeter of the proposed excavation. Spacing between soldier piles may vary from 4 or 5 ft to 10 or 15 ft.

In most cases, lagging is placed between the soldier piles. Lagging may consist of boards, but sometimes consists of precast concrete panels or cast-in-place concrete. If cast-in-place concrete is used, it may be the permanent final wall of the building.

In some cases, the soil is strong enough to bridge between the soldier piles. In these cases, lagging may not be used. If not, the soil between soldier piles may be protected from drying out. The protection may consist of chemical spray, plastic covering, or gunite. In addition, it may be desirable to protect against the possibility of a piece of soil falling off the vertical face. This may be accomplished by hanging wire mesh or fencing material

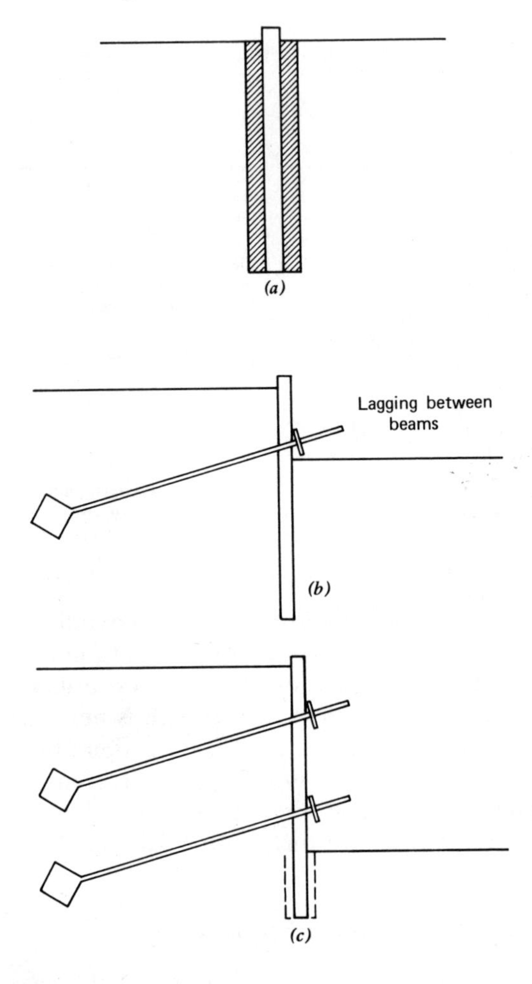

Fig. 16.4 Soldier pile system. (*a*) Step 1—Drill hole. Set soldier pile in hole. Backfill with concrete. Pile spacing generally about 6 to 8 ft on center. (*b*) Step 2—Excavate for first row of tiebacks. Prestress the tiebacks. Put wood lagging between the soldier beams. (*c*) Step 3—Excavate and install second row of tiebacks.

down the face and attaching it to the soldier beams. The soldier pile system is indicated in Fig. 16.4.

The soldier piles may be driven into place. In many cases, a hole is drilled, and the soldier piles are set into the hole. In this case, concrete may be poured to fasten the bottom of the pile into the soil below foundation level. The soil in the hole above foundation level may be backfilled with soil or with lean, low-strength concrete, easy to chip away later, when lagging must be placed.

Sometimes, holes are drilled for each soldier pile, using an auger, but the soil is not removed from the hole. Predrilling makes it easier to drive the soldier piles vertically.

16.3 Sheathing

Sheathing may consist of sheet piles, wood lagging between soldier piles, or boards held against the sides of trenches. Sheathing prevents soil from falling off the vertical bank, and also must push against the vertical face with sufficient force to prevent a slide-type failure. Such a failure surface is shown in Fig. 16.1.

Lagging usually is set behind the flanges of the soldier beams. In placing lagging, or wooden boards, it is necessary that the sheathing be tight against the soil. Soil in place has a certain strength. However, many soils, such as clayey and silty soils, lose strength if some slippage is permitted, and a failure plane develops. Therefore, sheathing must be pushed tight

Fig. 16.5 Continuous wall of poured-in-place concrete piles.

against the soil to keep slippage from starting. Then the soil strength helps to resist sliding. However, if the soil should slide even an inch or two into the gap behind loose sheathing, this soil has lost much of its natural strength and will impose greater forces on the sheathing. Where overexcavation results in the sheathing being loose, backfill or concrete or wood blocks should be placed behind the sheathing boards to press tightly against the soil face.

If a small amount of sliding occurs, even an inch, blocks of material may stretch and cause cracks to develop. This may damage adjacent pavements or structures. Much worse, however, the cracked soil becomes a sponge during rains, soaking up water rapidly and becoming much heavier and imposing greater loads on the sheathing.

Lagging should be separated, with cracks ½ to 1 in. wide between boards, so that water can drain out. In the past, burlap or "salt hay" (treated excelsior) was tucked into the cracks to pass water but prevent loss of soil (loss of ground). Now filter cloth placed behind the lagging or tucked into the cracks provides a better means of soil loss control.

Occasionally, vertical drilled and poured-in-place piles are constructed side-by-side in a row. They act as a combination of soldier piles and sheathing. Such an installation is shown in Fig. 16.5.

16.4 Bracing

The rakers or kickers are usually placed at an angle on the order of 30 to 40° from the horizontal (see Fig. 16.6). If the excavation is not too wide, crossexcavation bracing can be used.

The forces used to design the bracing should be calculated by a design engineer. A discussion of such calculations is presented in Ref. *37*.

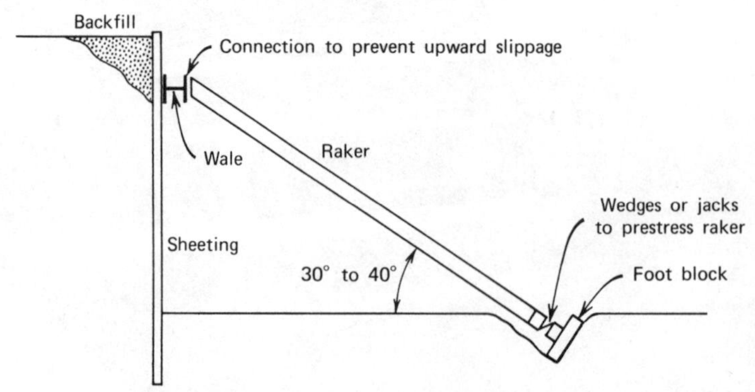

Fig. 16.6 Footblock may be wood, concrete, steel, piles, batter piles, or sheet piles. It is best to brace against a permanent interior foundation, if possible.

The rakers must be supported at the bottom by a footing, or foot block as shown in Fig. 16.6. The design of such footings is discussed in Chapter 18, Section 18.12.

16.5 Anchors

Anchors, or deadmen, usually are placed several feet below the ground surface. They may be individual blocks or can be a continuous beam. Generally, they can be designed by assuming that the passive resistance of the soil is approximately equal to the weight of the wedge of soil which would be pushed out if the soil failed. This "passive wedge" is indicated in Fig. 16.3. However, to take care of uncertainties regarding variations in soil conditions, it is general practice to use a factor of safety in the range of 1½ to 2 in designing the anchors. Therefore, the anchor geometry should be such that it would try to pull out a somewhat larger wedge of soil.

For deep excavations, it is sometimes convenient to put in anchors which are installed with drilling equipment. Such drilled-in anchors can extend down to and fasten into bedrock. As an alternative, the drilled-in anchors can extend far back into the soil and take their resistance by friction between the soil and a concrete cylinder. This is indicated in Fig. 16.7. A third alternative is to bell out a circular anchor, using a special belling tool. This scheme also is indicated in Fig. 16.7. This method is described in Ref. *37*, p. 375. The required length or size of anchors extending into rock or soil depends on the strength of the rock or soil. All anchors should extend behind a line which can be defined as the "line behind which failure is unlikely." If we examine the sketch in Fig. 16.7, the line along which failure is most likely might be Line *A*. A stability calculation would indicate that the factor of safety is 0.7.

Farther back, at line *B*, failure is less likely. At this location, the factor of safety might be 1.0.

Even farther back, at line *C*, the chance of failure is very unlikely. At this point, the factor of safety might be 1.5.

Depending on the degree of safety or factor of safety desired, a line such as line *C* would be picked as the area which is safe against sliding, or the "line behind which failure is unlikely." The anchor should be positioned in the soil or rock behind line *C*.

The pull-out strength of a cylinder of concrete can be estimated as the perimeter area of the concrete cyclinder times the shear strength of the soil along the cylinder. As an example, assume

Soil strength = 1000 lb/ft^2 average at depth of anchor.
Anchor diameter = 1 ft; surface area = 3 ft per foot of length.
Desired capacity = 50,000 lb.

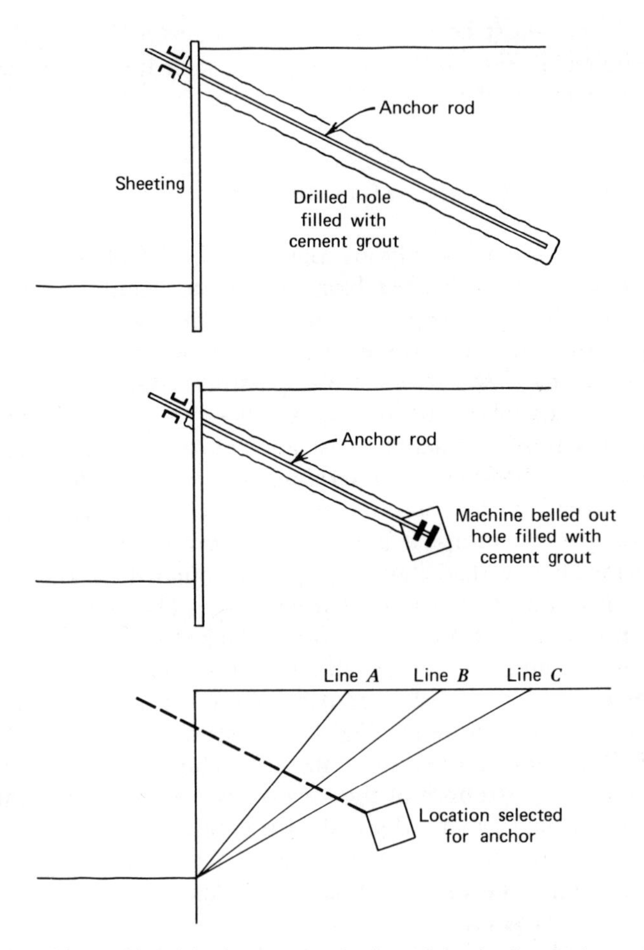

Fig. 16.7 Drilled tieback system's selection of depth of embedment.

Desired factor of safety = 1.5.
Required length = 25 ft.

For a belled anchor, the depth below grade greatly influences the bearing capacity. Assume that at the depth involved, the bearing capacity equals 20,000 lb/ft². Then, the required size of the anchor is

$$\text{area} = 3.75\ \text{ft}^2; \qquad \text{diameter} = 27\ \text{in.}$$

Pull-out tests are performed on most jobs to verify the capacity of the anchors.

Construction of anchors requires some skill, experience, and considerable cooperation from the soil. The soil characteristics required include the following:

1. The soil must be strong enough to stand in an open hole without caving.
2. The hole should be dry, although it is possible with some modification in construction methods to develop satisfactorily an anchor when the hole is below the water level.
3. Boulders or rock ledges, which make drilling difficult, deflecting the drill bit and causing a crooked hole, should be absent.

In most cases, anchors are considered to be temporary and useful in holding the excavation open during construction operations. In some cases, however, it may be desirable for the anchors to be permanent. In this case, the primary requirement is for the steel tie rod to be adequately covered by concrete or other protection to resist rusting and deterioration.

Anchor rods usually carry loads on the order of 30,000 to as much as 100,000 lb. The most common loads are on the order of 50,000 lb. The rods usually are on the order of 1 in. in diameter and are made of high-strength steel. The horizontal spacing between tie rods usually is on the order of 8 to 15 ft. Where deep excavations are required, the ties can be placed in rows, one below the other. The vertical spacing between ties is generally on the order of 6 to 8 ft. See Refs. *43* and *44*.

In the following section, a formula is given that may be used in determining the depth of embedment required to resist lateral loads where no constraint is provided at the ground surface by such things as rigid or ground surface pavement.

16.6 Freestanding Posts

Occasionally soldier piles are used to hold the shoring as free standing posts (see Fig. 16.8). The post develops resistance against leaning over by the

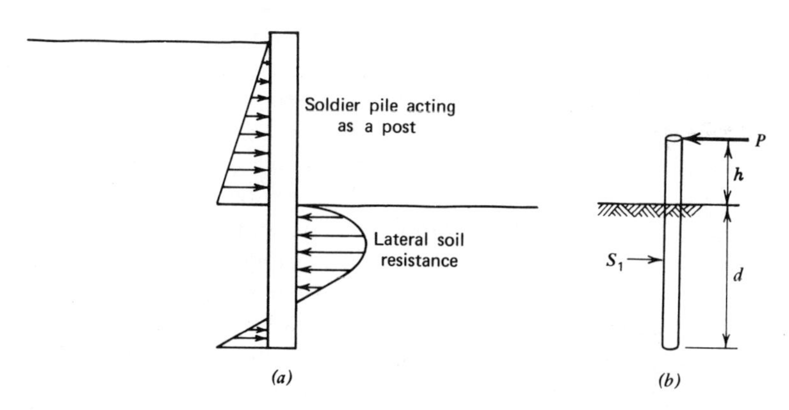

Fig. 16.8 Flagpole formula for calculating resistance to bending (from Uniform Building Code).

soil's passive resistance. There are several ways to calculate the strength against leaning over. Such formulas are described as flagpole formulas (see formula below and Fig. 16.8),

$$d = \frac{A}{2}\left(1 + \sqrt{1 + \frac{4.36h}{A}}\right)$$

where A = $2.34P/S_1b$.
P = Applied lateral force in pounds.
S_1 = Allowable lateral soil-bearing pressure based on a depth of one-third the depth of embedment.
b = Diameter of round post or footing or width of square or other shape post or footing (ft).
h = Distance in feet from ground surface to point of application of P.
d = Depth of embedment in earth in feet but not over 12 ft for purpose of computing lateral pressure.

16.7 Earth Pressures

16.7.1 Soil Types

The earth pressures behind shoring and bracing systems depend considerably on the type of soil to be retained. In the case of sand, the pressures are reasonably predictable and remain generally unchanged despite changes in weather, rainfall, and other conditions. By contrast, clay soils can be unpredictable. Many times, an excavation can be made to the planned depth vertically, with the clay soil standing without shoring. Shoring is needed primarily to avoid risk as a clay tends to dry out, and blocks of clay fall into the excavation. However, if an excavation in clay is left open for a period of several months, as may be necessary on a construction job, it may go through a variety of weather conditions such as rainfall, snow, or freezing.

After shoring has been installed, the primary concern is rainfall and saturation of the soil. Saturation can cause the soil to lose strength and to become heavier. This greatly increases the pressure on the shoring.

Assume that a shoring system to hold a clay bank was designed using an equivalent fluid pressure of 30 lb/ft^2 per foot of depth of soil. This same clay after saturation might lose most of its strength and weigh well over 100 lb/ft^3. The combined soil and water pressure might be on the order of 70 or 80 lb/ft^2 per foot of depth. This is more than twice the design pressure.

An even greater hazard can develop for shoring of soils which are expansive, such as adobe and gumbo clays. These soils expand when they become wet, and exert greater lateral pressures and tend to heave and to

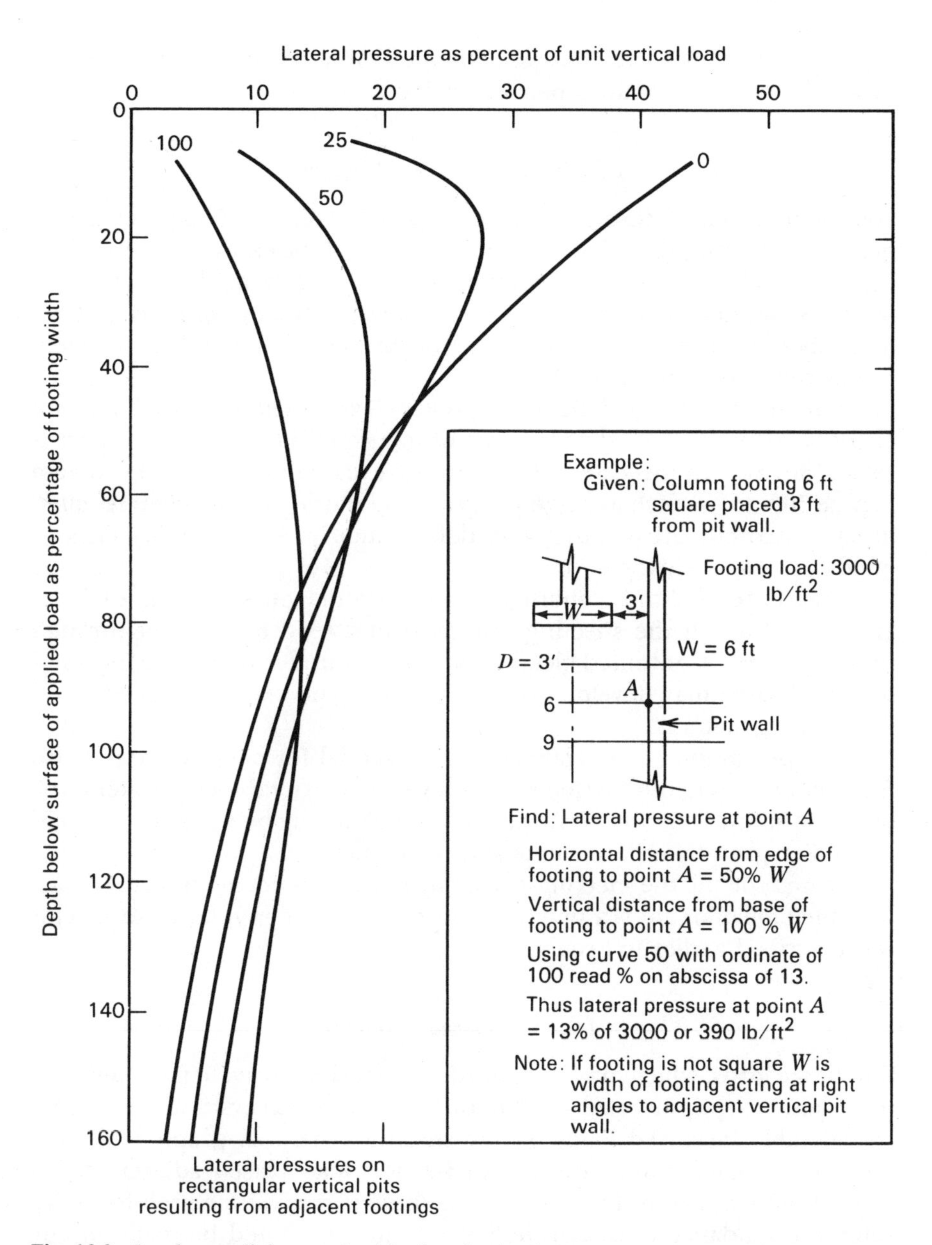

Lateral pressure as percent of unit vertical load
0
10
20
30
40
50
0
20
40
60
80
100
120
140
160
100
25
50
0
Depth below surface of applied load as percentage of footing width
Example:
Given: Column footing 6 ft square placed 3 ft from pit wall.
Footing load: 3000 lb/ft²
W
3′
W = 6 ft
D = 3′
A
6
9
Pit wall
Find: Lateral pressure at point A
Horizontal distance from edge of footing to point A = 50% W
Vertical distance from base of footing to point A = 100 % W
Using curve 50 with ordinate of 100 read % on abscissa of 13.
Thus lateral pressure at point A = 13% of 3000 or 390 lb/ft²
Note: If footing is not square W is width of footing acting at right angles to adjacent vertical pit wall.
Lateral pressures on rectangular vertical pits resulting from adjacent footings

Fig. 16.9 Load on wall due to wheel load or footing load. Numbers beside curves represent distance from edge of footing to vertical wall as a percentage of footing width, W.

expand laterally. The lateral pressure on the shoring system may be increased by several hundred pounds per square foot.

16.7.2 Construction Methods

Soils require some deformation to develop their strength. This deformation may be in the range of 0.1 to 1.0% of the height of the excavation. A flexible design, which presses consistently against the soil embankment but can permit some deformation to occur, may develop the optimum condition, sometimes called the "active" pressure. For this condition, the shoring system can be the minimum strength.

On the other hand, if the shoring system is very rigid and no deformation of the soil is permitted, the soil remains in a condition which sometimes is called "at rest." In this case, the soil pressure acting on the shoring system may be 50% to as much as 100% greater. Rigid shoring frequently is required in congested city areas; otherwise, deformation will crack utility lines or buildings near the excavation.

Sandy soil behind a shoring system may dry out. If cracks or holes are open through the sheeting, the sand may tend to flow out through these openings. Continued bleeding of soil may undermine large blocks of soil which then may develop greater instability and exert greater force on the sheeting system.

Proper drainage of retained soil is essential for permanent and also for temporary retaining systems. Where sheeting extends below water level, or where saturation may occur due to rainfall, it may be necessary to build in sand or gravel pervious filters and weepholes.

Constructing the sheeting or lagging so that it is not in uniform contact with the embankment is a potential cause for change in soil pressure. This was described earlier in Section 16.3.

16.7.3 Surcharge Loads

Surcharge loads, which may be applied from storing excavated soil, building materials, or equipment adjacent to the edge of the excavation, can impose considerably increased loads on the shoring system. A graph indicating increases in lateral pressure due to a wheel load placed adjacent to an excavation is shown in Fig. 16.9. While these curves can be used for quick reference guidance, a careful individual analysis should be made in any critical circumstances.

16.8 Slurry Trench

The slurry trench method has been used for many years in Europe. It was introduced in the United States in connection with major dam construction work, and more recently has been successfully used on several large commercial buildings in New York, San Francisco, and in connection with locks on the Tennessee-Tombigbee waterway.

Essentially, a trench is excavated in sections. Each section is kept continuously full of slurry. The slurry is basically a mixture of water and clay or mud. Frequently the "mud" is bentonite. The slurry may be made heavier by certain additives, such as Baroid. The hydrostatic pressure of the slurry is sufficient to support the banks and prevent failure. In addition, the mud cake on the walls of the excavation prevents the slurry water from entering the soil and causing it to soften.

After the trench has been dug to the desired depth, a concrete tremie pipe is lowered to the bottom of the trench. Concrete is poured on the bottom and gradually rises and fills the trench, forcing out the slurry. This method is described in more detail in Ref. *46*. One drawback to this method is that boulders, old wood piles, or other obstructions create problems in making the excavations.

16.9 Permanent Walls

In some cases, it is economical to make the temporary sheeting and lagging out of poured-in-place concrete. This concrete later becomes the permanent basement wall.

16.10 Trench Excavations

Deep trenches are required on more and more city streets for bigger and deeper utility lines. Many trenches are 50 ft deep. Trench excavations normally are crossbraced as they are dug.

Based on extensive measurements for deep excavations in clay soils, a lateral pressure diagram was developed which is considerably different from the common "equivalent fluid" system. This diagram is indicated in Fig. 16.10. A more detailed description of this method is given in Refs. *37* and *46*.

Since trenches may be of great length, sometimes many miles in length, the amount of shoring and bracing is a major expense. Occasionally, contractors build a test excavation and experiment with one or more methods of shoring and bracing to devise a workable and economical system.

Typical problems involved in projects include

1. Loss of ground adjacent to the bracing system.
2. Serious bending of walers, or bending of crossbracing or trench jacks.
3. Instability of bottom of excavation due to inadequate dewatering.

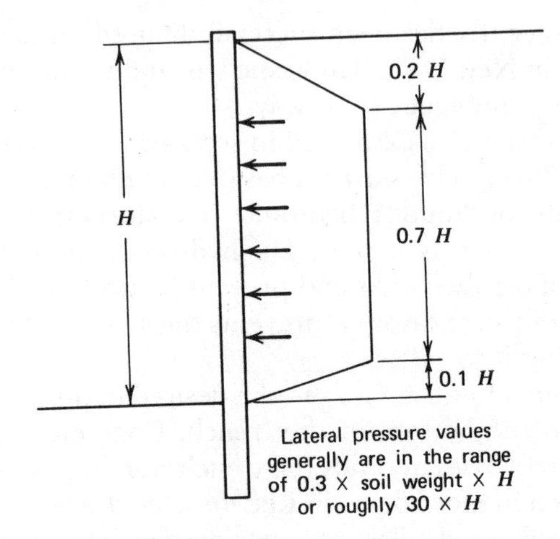

Fig. 16.10 Lateral pressure diagram for trench excavations in clay soils. The shape of the pressure diagram varies for various soils. Also, water levels higher than the bottom of the excavation impose additional loads.

4. Softness of bottom of excavation due to inadequate underdrainage, resulting in settlement of pipeline when trench is backfilled.
5. Adjacent building foundations, requiring underpinning, additional bracing, or stabilization of soil under the foundations.
6. Long sustained dewatering which may lower water under nearby structures supported on wood piles. These piles may rot in the "dry period" and lose so much strength that building settlements occur.
7. Undercutting of bedding planes or joints or fractures in rock, permitting a block of material to slide downward on a weak plane.

Studies have shown that trench cave-ins are a major cause of construction deaths. More than 100 per year are reported in some surveys. The accidents which result in trench cave-ins are mostly due to inadequate shoring or bracing, or as in about half the cases, due to too deep excavation without bracing.

17

Backfills

Probably the most important filling operation on most conventional building jobs, and one which is given the least attention, is backfilling. This includes backfilling around basement walls and backfills over utility lines and other buried service lines. These backfills frequently become important, because so many of them are done carelessly, resulting in broken water lines and also in settlements of floor slabs, sidewalks, streets, and highways. This leads to many lawsuits against contractors.

Backfills must be placed in restricted spaces, limiting the choice of equipment. Backfills should be compacted with mechanical equipment such as small sheepsfoot rollers, small vibrating rollers, or small rubber-tire wheel rollers.

Where space is more restricted, tamping equipment such as Ingersoll-Rand "simplex" or "triplex" tampers, Barco tampers, or Wacker tampers frequently are used.

In special circumstances, such as in clean sandy soils, backfills can be compacted by jetting. In most cases, however, jetting, ponding, or flooding produces a low-density fill. Such fills usually settle later.

17.1 Selection of Material

Backfills placed around structures are relatively small in volume. Therefore, the cost of material is small compared to the time of workers and equipment required to compact it into place. Many times, it is more economical to import good-quality material, which will compact easily, than to deal with on-site excavated soils which are difficult to compact.

17.2 Backfills behind Walls

Backfills behind walls generally are in deep and narrow slots. The biggest problem is to get workers and equipment down into the work area. Hand-operated tampers such as Ingersoll-Rand "powder puffs," Barco rammers, Wacker tampers, and similar machines generally are used. In most cases, engineers specify that the backfill be placed by mechanical methods. It takes energy and work to shove the soil down into a compact condition. Most soils will not compact by dropping into the hole or by flooding or jetting. Only very occasionally are the soil conditions "right" so that jetting will work (see Section 17.4).

In some cases, the lower part of a narrow excavation is filled by dumping in pea-gravel. As soon as the backfill is up high enough to permit normal working operations, the pea-gravel is tamped on top, and then soil backfill is placed in layers to complete the remainder of the backfill.

It is important that soil backfill, like general site grading fills, be placed in layers, as described in Chapter 22. Generally, the layers should be on the order of 6 to 8 in. thick, and each layer should be tamped before the next layer is placed. After the working area becomes about 5 or 6 ft wide, many kinds of small rolling compactors and vibrating compactors are available which are more efficient. Typical pieces of equipment include vibrating machines on the order of 3 ft wide. Manufacturers of such equipment are listed in Ref. *47*.

In addition, it is possible to use a piece of machinery on top of the bank with a boom extending down in the hole, with a vibrating tamper attached to the end of the boom. Such a machine can reach to depths on the order of 20 ft. It is more efficient than tamping with hand-operated small machines, and also increases safety, since a worker does not have to be at the bottom of the excavation.

After basement walls have been poured and stripped, the narrow slot between the back of the wall and the sloping face of the excavation can be a hazardous place to work. In most states, an area restricted by basement walls adjacent to an embankment is considered a trench, and the safety laws apply to workers in such locations in the same manner that they do to workers in pipeline trenches.

Lateral pressures develop on the basement wall as the fill is placed and compacted. Usually, there is a slight yielding of the wall. This yielding is usually small, probably less than 1⁄10 of 1% of the height of the wall. For a wall 20 ft deep, the yielding of the wall might be 1⁄4 in. Compacting of soil behind the wall may exert a fairly strong pressure on the wall. Therefore, temporary bracing frequently is placed to restrain the walls during placing of fill. An alternative is to place the interior floor slabs in position as cross bracing before the backfill is placed.

Soils engineers sometimes make the mistake of calculating lateral pressures on a retaining wall by considering the natural soil existing before start of construction. In construction, however, the natural soil usually is all ripped out. The wall is built and then some of the removed soil, or other soil, is backfilled behind the wall. The lateral pressures on the wall are generated primarily by the backfill. Therefore, the wall cannot be properly designed until the designer knows what the backfill will be. Generally, the engineer handles this problem by requiring that the natural soil be reused as backfill, or that a better soil be used, and that the soil be compacted to a specified density. If the natural soil is silt or clay, or is wet and sloppy, the specifications may require that other soil be used. Backfill generally is described in specifications by sieve sizes or by typical local names. Typical specifications for imported backfill soils are as follows:

> The soil shall be predominantly sand or sand and gravel, with not more than 20% passing the number 200 sieve.
>
> The fill shall be free of clods, wood, or masonry debris, or other deleterious material. The fill shall be compacted in layers not exceeding 8 in. in thickness, and shall be compacted to a density of 90% of the maximum, determined by the Modified AASHO Method of Compaction Testing, *28*.

Many walls are waterproofed, and drains are placed at the toe of the wall, as indicated in Fig. 17.1.

The backfill is placed in layers and compacted. However, the compacted fill may contain a layer of compacted silt or clay soil, as shown in Fig. 17.1. This layer would stop downward natural flow of water. The water would perch on the clay layer. It could not get down to the drain. Therefore, water would build up on top of the clay layer and might find a weakness in the waterproofing on the wall. In this case, the water would flow through the wall.

For basements extending below groundwater, drainage is necessary during construction. Usually, well points, deep wells, perimeter trenches and sumps, or other systems are installed to draw down the water level. A

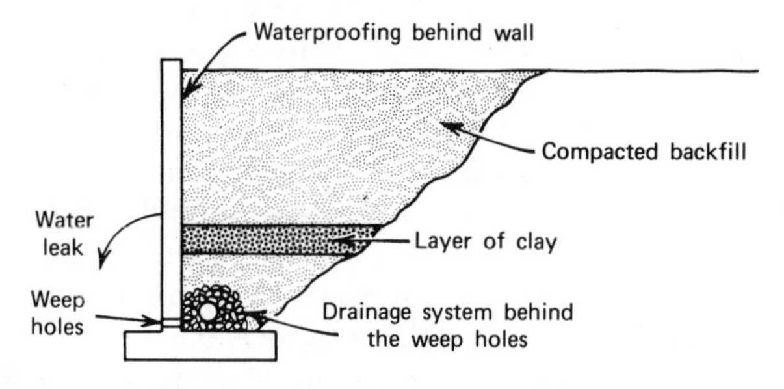

Fig. 17.1 Failure of good drainage system due to improper back filling.

tendency for water to collect at the sides of the excavation is a problem in attempting to put in a good compacted backfill. Therefore, one approach is to use pea-gravel or clean sand for the first few feet of the backfill. If necessary, small sumps can be installed temporarily to permit pumping out excess water.

17.3 Backfill for Large Culverts

Large corrugated metal culverts are frequently used under roadways. The corrugated metal is quite flexible and is not sufficiently strong to act as a bridge. Therefore, it depends to a great degree on the strength of the soil backfill surrounding the culvert for stability. Detailed construction procedures for placing such fills are outlined in manuals, such as Ref. *48*.

17.4 Backfills in Utility Trenches

Utility trenches are constructed inside building sites, from the building sites to streets, through streets, and cross-country. There are many theories about pipeline construction and backfilling.

The bedding and support for pipes are extremely important. In most cases, the pipes themselves are designed structurally to carry the overburden soil loads only, with some help from the supporting soils. In these cases, bedding or shaping of the bottom of the trench to fit the contours of the pipe or other schemes for providing support in the lower one-third or one-half of the pipe are important to prevent the pipe from squashing down and cracking. Several typical cross sections for providing bedding for pipelines are shown in Fig. 17.2. Detailed discussions of culvert or pipeline bedding are contained in Refs. *48* and *49*.

In some cases, it is considered more economical to put additional money into the pipe itself. It is made structurally strong enough to carry the full overburden soil pressure even though bearing only on a flat, hard surface. In this way, special bedding procedures and special care and procedures in placing backfill around and over the pipe can be eliminated in cases where no overlying structure must be supported.

Pipeline excavations can be cut vertically or with sloping sides. Vertical excavations require removal of less material and can be cut with trenching machines. However, most vertical-sided trenches must be braced or shored. In wide open country, it may be cheaper to slope the sides of trenches to eliminate bracing and shoring. The shape of the excavation can have a considerable effect on the load that the backfill imposes on the pipe, and sometimes specifications will require a particular shape.

Trench bracing is specified, with typical designs, by federal and state safety codes. However, there is considerable leeway within the codes, since

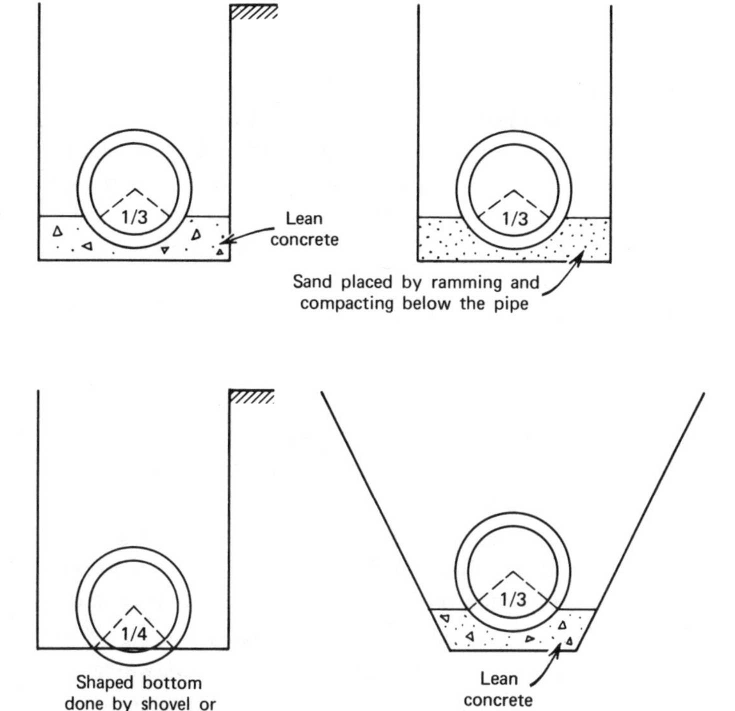

Fig. 17.2 Pipe bedding. For more information, see Refs. *48* and *49*.

the codes have fairly unspecific definitions for soil conditions. There is a lot of ground between "hard" and "soft." Therefore, the contractor has considerable leeway in selecting a shoring system and must make several decisions.

For deep trenches, some soils engineering information can be helpful in preparing trench designs. Alternately, experiments can be made. Test trenches can be dug and shored using a particular design, to see if it will work. If trouble develops, then a revised design can be tried. Lateral pressures on shoring in the dry may be on the order of 20 to 30 lb/ft of depth. However, below water level, these pressures can become three times greater, such as 70 to 80 lb/ft of depth.

Backfill in pipelines has traditionally been placed by dumping in the soil, and then flooding or jetting the soil with water to cause settlement. Over the years, however, it has been found that such backfills generally settle and subside. Where they underlie roadways, the roads continually require repairs, and the roads are rough and unsatisfactory to motorists.

When can flooding or jetting be used? Generally, it has been found that clean sand soils will compact moderately well by flooding and jetting.

If the soil at the bottom of the trench is free-draining sand and the water level is below the bottom of the trench, the jetting water will flow downward through the fill soils and out of the bottom of the trench. Generally, downward drainage results in reasonably good compaction of sand backfill. Tests on sand backfill under these conditions usually indicate compaction of 85 to 90%, based on modified AASHO. However, for silt or clay backfill soils or sand containing some silt or clay, jetting is not very effective. The backfill usually remains soft and sloppy for some time and test densities are usually below 85% compaction.

In St. Louis, a committee of the American Society of Civil Engineers performed a 5-year study of backfilling practices. Their report contained the following conclusions:

1. The mechanical (hand) tamper is relatively ineffective in obtaining consistently satisfactory dry densities in cohesive soil backfills for use under sidewalks, pavements, and lightly loaded structures.
2. The mobile trench compactor (under certain conditions and with certain limitations) can produce consistently satisfactory dry densities in cohesive soil backfills for use under pavements and lightly loaded structures.
3. The jetting method of compacting cohesive soils and trenches does not produce satisfactory dry densities for use under sidewalks, pavements, and lightly loaded structures within a reasonable period of time.

For backfills in streets, it is common but poor practice for the bottom several feet of fill to be placed in fairly thick layers, such as 2 ft thick, with only a moderate amount of compaction. The top 2 to 3 ft underneath the pavement are generally required to be compacted to 90%. As a result, the pavement patch must be considered temporary, and repeated patching and repairs will be required in the future as settlement occurs.

In cross-country pipelines, the backfill generally is thrown in loose, and the top is wheel-rolled. However, compaction is required for road crossings. For crossings of railroad tracks or freeways, tunneling with a horizontal auger drill is common.

Where pipelines underlie foundations of structures or other facilities which cannot permit settlements, it is necessary for fills to be placed in thin layers, 6 to 8 in. thick, and for each layer to be compacted with mechanical equipment. For these conditions, mechanical equipment is required even for sandy soils. Such backfills usually are tested by the engineer to assure proper compaction.

17.5 Pipelines

It is extremely difficult to build pipelines under submerged conditions. In addition, there may be difficulty in holding a pipeline down. Frequently, pipelines and buried tanks, when empty, have popped out of the ground because of high water level in the ground. For this reason, the commonly used well point system was developed. It is by far the most satisfactory system for temporarily dewatering pipeline alignments, for stabilizing the excavation, and for permitting backfill to be placed and compacted.

Sometimes pipelines are constructed by placing a gravel base below the pipe invert and submersible pumps in the gravel to pull out the water. This works well in many soils, but in some cases the upward movement of water through the soil can cause a soft or even a quick condition on the bottom of the trench, which results in wavy inverts and lawsuits.

Frequently, pipelines, such as gas pipelines of large diameter, are emplaced by making wide excavations, securing the pipeline in position by use of screwed-in anchors, and then replacing the backfill loosely. Releveling and grading are done to compensate for large settlements of the backfill. This method is satisfactory on cross-country pipelines, but not under pavements.

A wide variety of equipment is used for compacting backfill in trenches. At the bottom of the trench and around the haunch of the pipe, soil may be shoved into place by shovel or by ramming with sticks, or compacted by small vibrators or hand-operated tampers. As the fill comes up to the top of the pipe, small vibrators can be used. After the fill is about 1 to 2 ft over the top of the pipeline, it is customary to use rubber-tire rollers, small tractors, or vibrating drum rollers. After the fill is up to within 2 or 3 ft of the surface, larger equipment such as sheepsfoot rollers or truck wheel rolling is common.

In vertical trenches, the lower portion of the fill is compacted by small hand tools, and only the top 2 or 3 ft of fill can be compacted by large mechanical equipment.

It is difficult to moisten or otherwise condition the fill soil "in place" in the bottom of a trench. Therefore, if the soil must be dried, moisture added, or other changes made, these changes should be made on the top before the material is dumped into the trench. On top, there is room for using trucks with spray bars for moistening the soils, and for using scarifying tools for mixing the soil.

Rocks can be a problem in the backfill, since they puncture the protective wrapping. While the excavated soil is on top, soil free from rocks can be selected and placed around the pipe. Otherwise, the pipe wrapping must be thick or protected by some method.

18

Foundations

18.1 Spread Footings

Spread footings generally consist of a pad of concrete placed at a shallow distance below the ground surface to support a building column. Spread footings are the most common type of building foundation.

Typical average bearing values for use in the design of spread footings are listed in Table 18.1. These values are typical of bearing values listed in many building codes. Generally, these bearing values are considered conservative. Usually, higher bearing values can be determined as a result of sampling and testing of the soils. Most building departments will accept higher loads based on laboratory tests and soils engineering recommendations.

Sometimes bearing values are presented in graphical form. A typical bearing value graph is presented in Fig. 18.1.

The soil below the footings should be as firm as the soil on which the footings rest, down to a depth of at least 1.5 times the footing width. Also, the foundation soil should be the same classification as shown on the plans. Soil conditions frequently change appreciably from one footing to the next. In general, a footing should bear on uniform soil. If there is a marked change, contact the soils engineer.

18.2 Footings on Sloping Ground

Usually, footings are stepped on sloping ground in staircase fashion. Occasionally, where the ground surface slope is relatively flat, foundations are poured to conform to the slope.

Where one footing will be near another and at a higher elevation, it is common practice to restrict the placement so that a plane drawn from the lower corner of high footing, at an angle of 45° from the horizontal, will not intersect the lower footing. This concept is shown in Fig. 18.2. Obviously, the lower footing should be built and backfilled first. In some cases, however, it will be necessary to dig down below an adjacent footing to install a new footing at a lower elevation. The slope angle above may not always be safe for this condition. If, for instance, the soil is sand with a friction angle of 30°, the stable slope of 30° would mean that the upper footing would fail. Consult the soils engineer on methods of handling this problem.

If it becomes necessary to place a footing closer than the 45° line, the bearing value for the upper footing may have to be limited, and it may be expected that the lower footing would experience somewhat larger settlements, since it will carry some of the load of the upper footing. It may be better to place the upper footing at greater depth. Also, the upper footing could be moved back and the footing-to-column connection designed for eccentricity.

There is another problem—dealing with foundations on sedimentary rock or stratified soil. This condition is shown in Fig. 18.2. Even though the soil or rock is firm, each bedding plane on the right side of the trench is a plane of possible sliding. The wedge *ABC* cannot be used for foundation support. The footing at the right must be kept beyond point *C*. By contrast, the footing at the left can be placed closer to the trench.

18.3 Continuous Footings

Continuous footings, such as wall footings, strap footings, and girder foundations are spread footings made long enough to carry a wall or a row of several columns.

The excavation for such long footings can sometimes be done with trench diggers, which can be a speedy and economical excavation system. Such excavations are trimmed by hand to the neat size of a proposed continuous footing, and concrete is poured without the use of side forms. Elimination of carpenter work for forming can save time and money. These foundtions are always reinforced.

Continuous footings have the advantage of being able to bridge or span local soft areas. Therefore, they may offer more uniform support for a long wall or row of columns than individual footings would provide.

18.4 Mat Foundations

Mat foundations frequently are used under very heavy structures. They are continuous in both directions and therefore occupy a large area. Mat

Table 18.1 Typical Bearing Values for Design of Spread Footings

Class of Material	Minimum Depth of Footing below Adjacent Virgin Ground (ft)	Value Permissible if Footing Is at Minimum Depth, lb/ft^2	Increase in Value for Each Foot of Depth That Footing Is below Minimum Depth, lb/ft^2	Maximum Value lb/ft^2
1	2	3	4	5
Rock	0	20% of ultimate crushing strength	0	20% of ultimate crushing strength
Compact coarse sand	1	1500[a]	300[a]	8000
Compact fine sand	1	1000[a]	200[a]	8000
Loose sand	2	500[a]	100[a]	3000
Hard clay or sandy clay	1	4000	800	8000
Medium-stiff clay or sandy clay	1	2000	200	6000
Soft sandy clay or clay	2	1000	50	2000
Expansive soils	1′6″	1000[b]	50	
Compact inorganic sand and silt mixtures	1	1000	200	4000
Loose inorganic sand silt mixtures	2	500	100	1000
Loose organic sand and silt mixtures and muck or bay mud	0	0	0	0

[a] These values are for footings 1 ft in width and may be increased in direct proportion to the width of footing to a maximum of three times the designated value.

[b] For depths greater than 8 ft use values given for clay of comparable consistency.

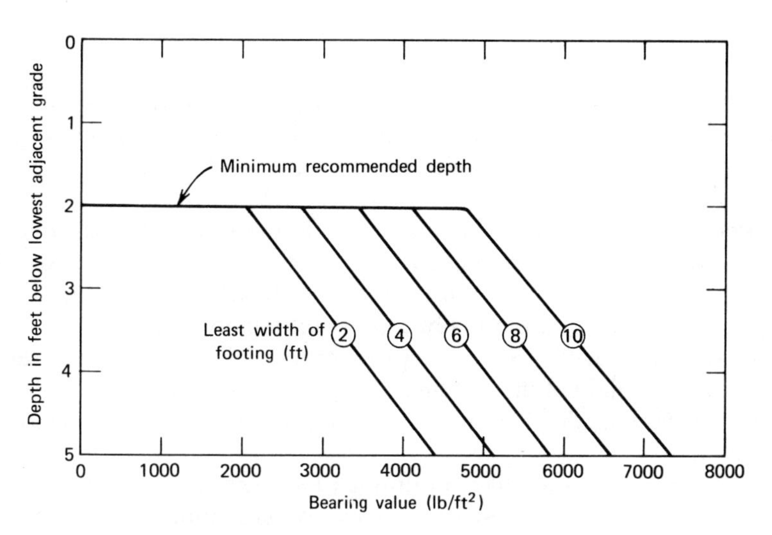

Fig. 18.1 Bearing value graph.

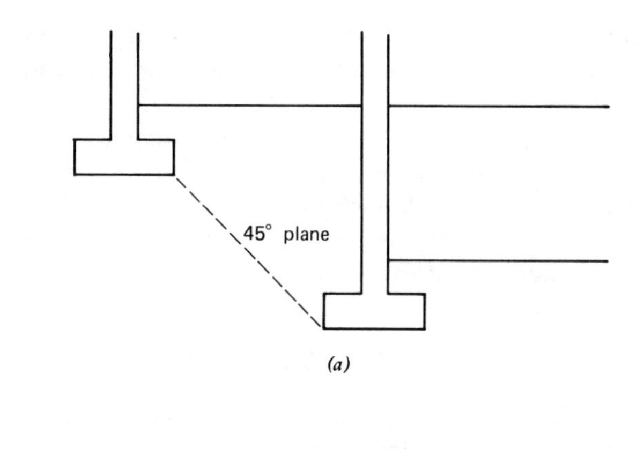

(a)

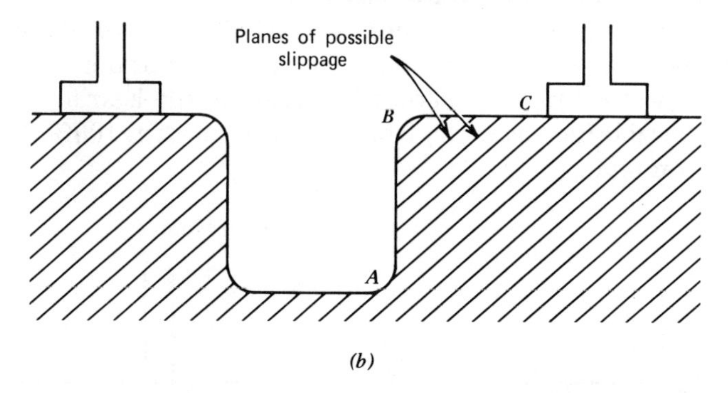

(b)

Fig. 18.2 (*a*) Limitations affecting adjacent stepped footings. (*b*) Potential failure along bedding planes of rock into a trench affects location of adjacent spread footings.

foundations usually are thick, involve a large volume of concrete and a considerable weight of reinforcing steel, and are expensive. They can be an advantage where there is upward hydrostatic pressure due to high water level.

Since a mat foundation, with the sidewalls of the building, acts somewhat like a barge, it has some advantages. The weight of soil excavated before pouring the mat foundation can be deducted from the total weight of the building, which may substantially reduce settlement. In some cases, mat foundations are placed at such a depth that the weight of soil removed will equal the gross weight of the building. In this case, future building settlements usually are very small, since no new load is placed on the soil below the mat foundation.

Mat foundations act like continuous footings in both directions, and therefore can bridge over soft spots. This results in more uniform support for building columns.

18.5 Slab Foundations

Slab foundations frequently are used under one-story and two-story light structures such as houses, school buildings, and light industrial and commercial buildings.

Usually, the edge of the slab is thickened to form a perimeter footing. In addition, ribs or pads may be placed as thickened parts of the slab to support walls or footings. Such a foundation is shown in Fig. 18.3.

This design is primarily used for economy. It is not suitable in areas of deep freezing, high water levels, or expansive soils.

18.6 Temporary Loads

The bearing values given in Fig. 18.1 apply to static dead loads, static live loads, and frequently applied live loads.

Temporary loads may be applied by wind forces, earthquake lateral forces, impact forces, or other loads of very short duration. For these circumstances, it is common to allow an increase in the bearing pressure on the soil. Allowable increases vary from 10% to as much as 100%. An increase of 33% is common.

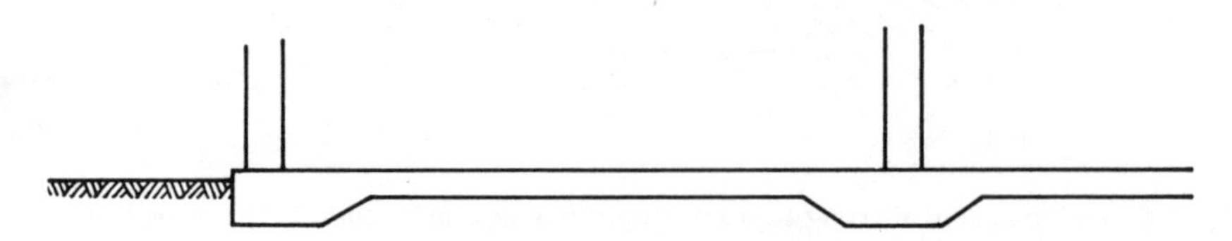

Fig. 18.3 Slab foundation.

Temporary loads may be applied to footings during construction. For instance, tilt-up walls frequently are placed on footing pads at column locations, and then the continuous footing under the tilt-up wall is poured.

In such a case, the footing may be loaded to a much higher bearing pressure than its design pressure when the building is completed.

Generally, there is reserve strength to permit overloading the foundation. This is sometimes called the factor of safety. Factors of safety frequently range from 1.5 to 4.0.

If temporary loads are planned which will considerably exceed the design bearing value, the plan should be reviewed with the structural designer of the building to make sure that the reserve strength of the foundation will not be exceeded. Sometimes foundations are failed by temporary construction loadings.

18.7 Compacted Soils Foundation

Many soils are deposited naturally in a loose state, so that they provide poor support for spread foundations. Therefore, pile foundations or other expensive foundations are used frequently.

In many such cases, it is more economical to excavate the soft or loose soils. Sometimes these soils can be reused for constructing compacted fills; in other cases, imported soils are necessary to construct a suitable soil fill. These soils are used to backfill the excavation. Soil fills can be engineered and constructed to support desired foundation loads. Such fills become part of the foundation structure system, and sometimes are referred to as "structural fills."

In one case of this type, the foundation for a power plant was to bear on a layer 20 ft thick of loose silty fine sand. It was found to be economical to remove the loose soil. After removal, the soil was recompacted into the excavation. It shrank from a thickness of 20 ft to 15 ft. Additional soil was imported to make up the shrinkage. The foundations for the power plant were placed on this "structural fill" (see Ref. *50*).

Occasionally, soils are compacted in place to increase their bearing capacity for support of foundations. Methods of compaction in place are described in Chapter 22.

18.8 Footing Locations

In early stages of construction, some confusion on the site may cause one or more foundations to be poured out of position. A question then develops regarding usability of the footings.

In some cases, the footings were lifted out by crane and reset at the proper locations. In one case, a number of footings were 2 ft out of position.

Holes were drilled into the footings. Bolts were grouted into the holes, and the footings were lifted out. The bottoms of the footings were hosed off. New footing excavations were cut, extending 4 in. over depth. Four in. of fresh concrete was poured, and the footings were set on the fresh concrete.

Sliding footings laterally a few inches by use of jacks or heavy construction equipment has been given serious consideration in more than one instance, but the writers do not know of an actual case. This appears to involve some risk, because the soils under the footings are disturbed and loosened, which encourages settlement.

It is possible that the column connection to the footing could take some bending moment, in which case the offset footing still could be used, in its offset position. This must be reviewed with the structural engineer.

Where footing excavations are cut neat into the soil and side forms are not used, foundations occasionally extend out farther on one side than planned. This would appear to result in an eccentric foundation. However, so long as the foundation is as large or larger than planned in the other three directions, no case is known where such possible eccentricity has created a problem.

18.9 Drying or Saturation

The soils in the bottom of a foundation excavation may be altered considerably by drying or by saturation.

Drying should be restricted and can be controlled by frequent sprinkling of the soils, or covering with plastic, canvas, loose dirt, or straw.

Soils which are seriously dried and which shrink and crack badly may be expected to regain their moisture content after the footing and the building have been completed. Regaining moisture may cause the soils to swell and heave, possibly lifting some foundations or the floor slab.

Foundation excavations frequently become saturated during heavy rains, since water collects at this low area. Generally, the most satisfactory solution is to overexcavate to remove the soft soil.

If the reinforcing steel already has been placed, other experiments may be worth trying. Probings could be made to determine the thickness of softened soil and the likely increase in settlements. For flexible structures, the anticipated additional settlement may not be too serious a problem.

In a few cases, heaters and blowers have been used to cause rapid drying of saturated soils.

18.10 Overexcavation

Many foundation excavations are made with large excavating machinery. Large machines working in small restricted spaces sometime are difficult to control, and overexcavation may occur.

In one such case, foundation excavations were made, steel was placed, and the concrete truck was there ready to pour concrete. Inspection of the bottoms of the excavations included probing with a steel rod. In almost all excavations, the rod could be pushed easily 6 to 12 in. into the bottom soils. The bottom of the excavation was level. It was found that the operator had dug deeper as much as 12 in. Then he put loose soil back in the excavation to level the bottom of the excavation.

Many cases have been noted in which the soils at the bottom of foundation excavations had been disturbed as a result of accidental overexcavation. Replacement of the soil was done hurriedly to hide the mistake. Therefore, the replaced soils were not adequately compacted, but were looser than the natural soils.

Generally, it is cheapest to accept overexcavation. The overexcavation can be made up with lean concrete. Lean concrete is simple to pour and may be appreciably less expensive than attempting to recompact soils in a small area. Lean concrete backfill generally is considered part of the soil foundation, not an extension of the concrete foundation itself.

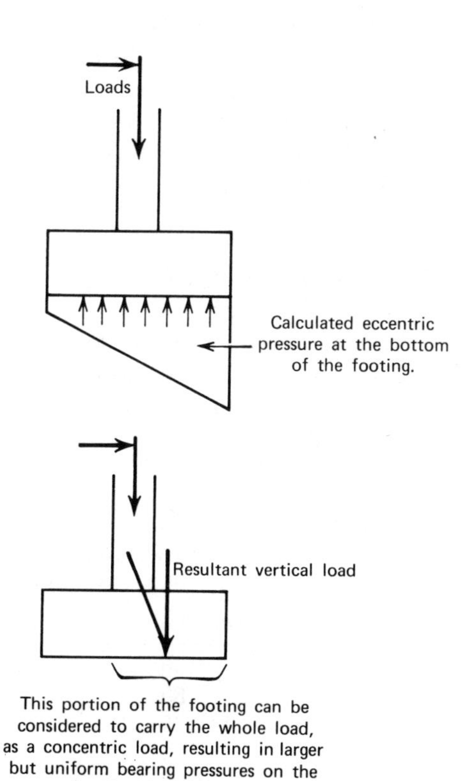

Fig. 18.4 Distribution of bearing pressure under eccentrically loaded footing.

Frequently, a foundation bottom is irregular due to problems in excavating the soil. A natural tendency is to "level up" the bottom of the excavation. It looks better and is easier to set the steel. Leveling up is done with soil or gravel. The leveling material should be compacted and tested. However, most soils engineers agree that it is better to leave the bottom rough, in undisturbed soil, and to pour extra concrete to fill in the low areas. The extra concrete can be poured first, if this makes it easier to set the reinforcing steel.

18.11 Eccentric Loads

Overturning loads, resulting in eccentric forces on foundations, are common for structures subject to wind or earthquake loads, or for retaining walls.

The common method of analyzing the foundation pressures has been to convert eccentricity into a triangular-shaped soil pressure. This is indicated in Fig. 18.4.

Usually, the designer restricts the maximum edge pressure to the allowable design bearing pressure, or to the design bearing pressure plus an allowable increase for temporary loads.

An alternate solution also is shown. In this case, it is assumed that eccentricity places the center of force at a new location. One end of the footing can be "theoretically trimmed off" so that the resulting remaining portion of the foundation is concentric. In this case, the smaller foundation can be designed using the full design bearing pressures as described previously.

18.12 Inclined Footings

Inclined footings sometimes are used as foot blocks for bracing, as tieback anchors, and as anchor blocks for pipeline bends (see Fig. 18.5). Such footings have a lower bearing capacity than vertically loaded footings at the same average depth. A rough guide for estimating the reduction in bearing capacity is given below.

Inclination of Load from Vertical (deg)	Usable Bearing Value as a Percentage of Vertical Bearing Value	
	For Clay (%)	For Sand (%)
0	100	100
30	80	40
45	70	25
60	60	15
90	50	10

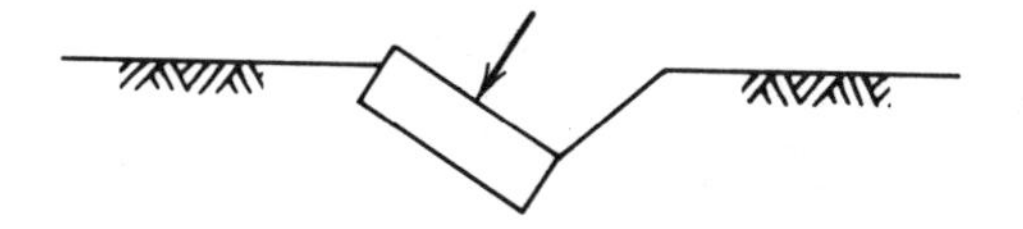

Fig. 18.5 Typical inclined footing. See Ref. 37.

18.13 Foundations on Expansive Soils

Several constructors of housing tracts have been hurt financially as a result of constructing numbers of houses on expansive soils. These soils expand when they get wet, and shrink when they dry out. Volume changes may be 5 to 10% or more (see Chapter 11, Section 11.11, Chapter 12, Section 12.11, and Fig. 18.6).

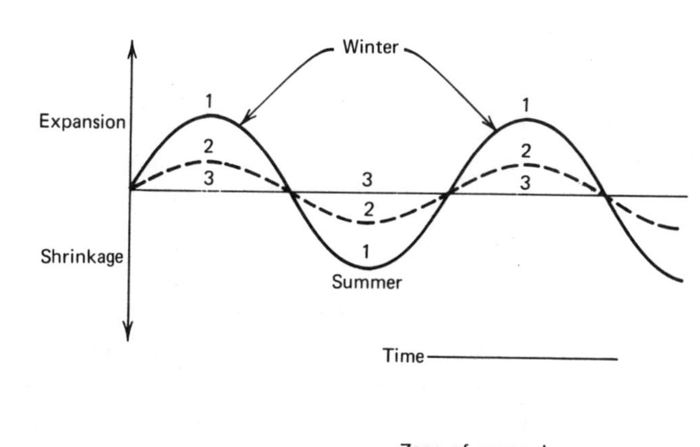

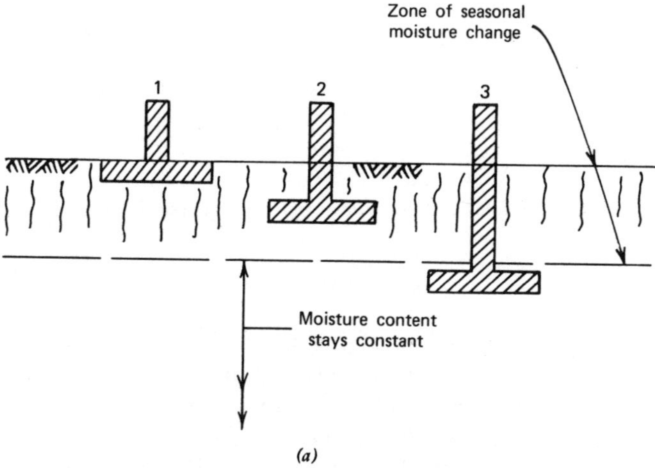

(a)

Fig. 18.6 (*a*) Seasonal behavior of expansive soils.

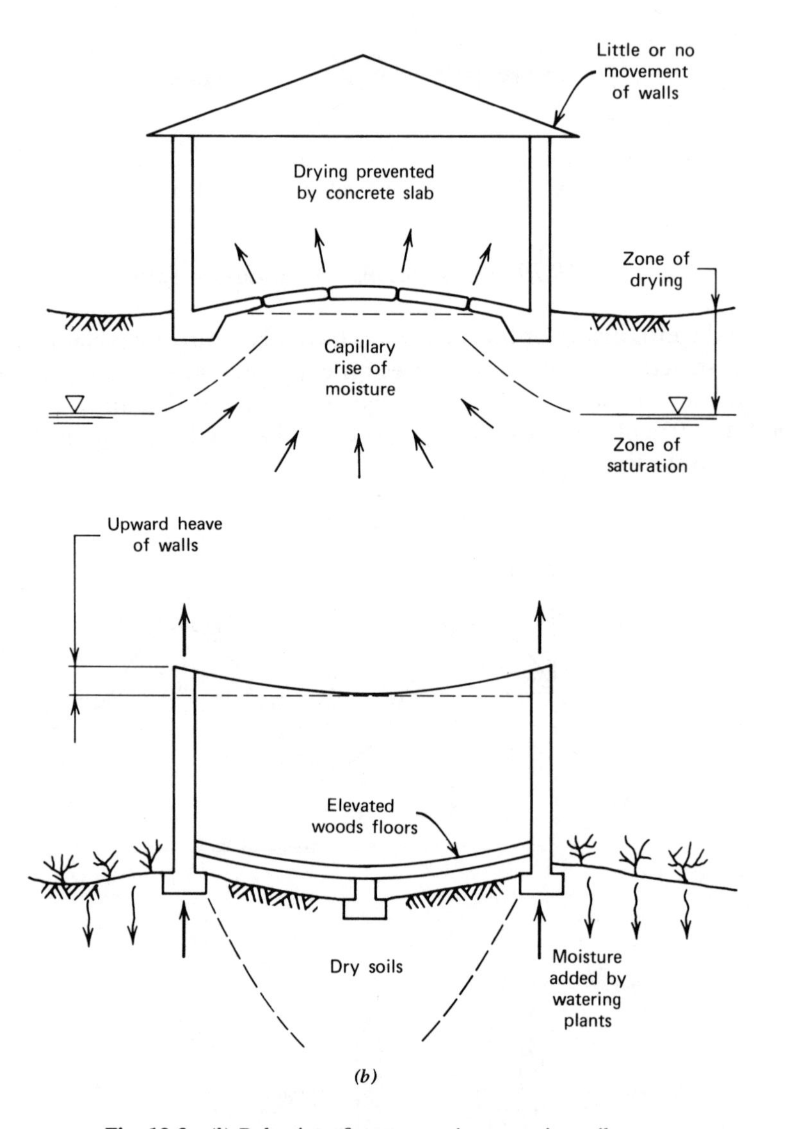

Fig. 18.6 (*b*) Behavior of structures in expansive soils.

The first problem is identifying the soil. Sometimes a clue can be found by examination of the site. Soils that are expansive usually are hard and cracked when they are dry. Cracks in the ground may be ½ to 1 in. wide and several feet deep. The hard clods of soil are like brick and cannot be broken easily. When wet, the soil is plastic, like modeling clay.

At least four approaches are used to solve the problem:

1. Keep the moisture content constant.
2. Place foundations below the depth of moisture change.
3. Treat the expansive soil with chemicals such as lime to stabilize it.
4. Excavate and remove the expansive soil in building areas. Replace with stable soil (see Fig. 18.6*a* and *b*).

18.14 Foundations on Hot Soils

Chemicals in some soils can corrode steel and deteriorate concrete. They are called hot soils (see Chapter 11, Section 11.13, Chapter 12, Section 12.10, and Chapter 14, Section 14.9). If laboratory tests or field resistivity tests indicate such conditions, special precautions may be necessary to protect the concrete and the reinforcing steel. Based on experience, tables have been prepared showing percentages of chemicals which could cause problems. Based on Ref. *51*, an abbreviated table showing the amounts of chemicals which could cause problems is as follows:

Percent SO_3 Contained in			
Groundwater	Clay Soil	Seriousness	Recommendation
0–0.03	0–0.2	No special problem	No special measures
0.03–0.1	0.2–0.5	Some problems especially in thin concrete walls	Use sulphate resistant Portland cement concrete (type II)
Over 0.1	Over 0.5	Serious problems	Use high alumina or super-sulphate cement (type V)

A typical solution in high sulphate or high chloride soils is to use more resistant cement, such as type V, to resist the hot soils. Also, steel should be protected by a thicker cover of concrete, such as 4-in. cover minimum.

19

Settlement

19.1 General

During construction, as building column loads are placed on foundations, the foundations will settle.

If the foundations are on extremely hard soil or rock, the settlements may be very small. However, if they are on ordinary valley soils, settlement may be a fraction of an inch or several inches. Settlements of ½ to 1 in. are common.

A substantial amount of the settlement may occur during construction. In other cases, settlements occur very slowly, and for many years after construction is complete (see Ref. *52*).

19.2 Calculated Settlements

A system of laboratory tests and calculations has been devised to estimate settlements for a proposed foundation.

Under Chapter 12, Section 12.6, consolidation tests were discussed. These tests measure the compression of soil under load, and the speed at which compression occurs.

When load is placed on a foundation, it is transferred downward to each of the soil layers underlying the foundation. This pressure distribution is indicated in Fig. 19.1. Immediately under the footing, the soil pressure is increased to approximately the full bearing pressure load. At depths of 10 or 20 ft below the ground surface, the load is spread out, and the increase in pressure is small.

To estimate settlements, the soil underlying a proposed foundation is divided into layers. The settlement of each layer is estimated; then these numbers are added together for the total settlement.

The natural soil pressure curve in Fig. 19.1 indicates that the soil pressure at a depth of 5 ft is approximately 500 lb/ft^2 since the soil weighs approximately 100 lb/ft^3. The new foundation will have a bearing pressure of 3000 lb/ft^2, due to dead and other static real loads. Momentary loads due to wind, seismic, and other infrequent and short-term loadings are deducted generally, since they have relatively little effect on settlement. The soil pressure is distributed through the soil, and therefore increased pressure at a depth of 5 ft is not 3000 lb/ft^2, but approximately 2100 lb/ft^2.

Since the normal pressure at a depth of 5 ft is 500 lb/ft^2, this pressure will be increased to 2600 lb/ft^2 as the result of the footing. Based on the laboratory consolidation test, it was found that a sample obtained from a depth of 5 ft consolidated very little when a load of 500 lb/ft^2 was applied. This is reasonable, since the soil already had been loaded to 500 lb/ft^2, and already had consolidated. The very small consolidation is taken as the starting point for measurement. Continuing the consolidation test by applying heavier loads, it is found that the soil consolidates an additional 1% when the full load of 2600 lb/ft^2 has been applied to the soil sample.

Assuming this layer is 5 ft thick, the total compression of the layer is 0.6 in. (60 in. × 1% = 0.6 in.).

Similar calculations can be made for all of the other layers under the foundation. The total is the expected settlement of the footing, probably in the range of 1.5 in.

Some typical values of estimated settlements for foundations on various types of soils are indicated in Table 19.1.

19.3 Measured Settlements

Settlements frequently are measured. The measurements can be made most easily and accurately if bench marks are set in columns during early phases of construction.

If no such bench marks were set, settlement surveys are compared back to the "as-built" elevations for foundations or supported floor slabs. These records usually are much less accurate.

19.4 Settlement during Construction

Certain soils, such as sands and free-draining materials, settle quickly under load. These settlements may occur almost entirely during the construction period. Therefore, practically no settlements will occur after the building is completed.

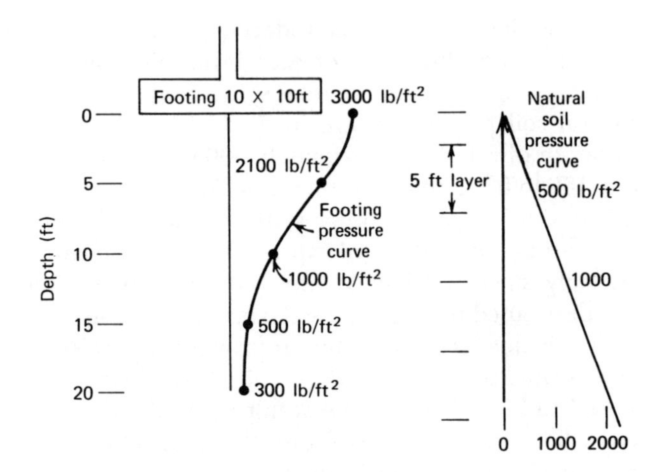

Fig. 19.1 Pressure distribution through soil.

By contrast, silt and clay soils are slow draining. Therefore, settlements will occur during construction but also will continue for many years after construction is completed.

During consolidation testing in the laboratory, the speed at which the test sample compresses can be measured. This is an indicator of the speed of settlement to be expected for a foundation.

A method of calculation has been set up to estimate the time required for most of the settlement to occur. Drainage of a layer of soil depends on its rate of drainage and the thickness of the layer. When a sample of a particular soil layer has been tested, the time required for consolidation of the soil layer can be estimated by comparing the thickness of the test sample with the thickness of the soil layer in the ground.

Table 19.1 Typical Settlement Values for a Column Load of 300,000 lb

Soil Type	Bearing Pressure (lb/ft²)	Footing Size (ft)	Settlement (in.)
Hard clay	10,000	5½ × 5½	0.5
Compact sand	10,000	5½ × 5½	0.3
Mod. firm clay	3,000	10 × 10	1.5
Mod. compact sand	3,000	10 × 10	0.8
Soft clay	1,500	14 × 14	3.0
Loose sand	1,500	14 × 14	1.5

Test Sample	Soil Layer
Thickness = 1 in.	Thickness = 10 ft.
Speed of settlement = 30 min.	One-half thickness = 60 in.
Note: Drainage from top and bottom applies to the test sample and to the soil layer. Thus, drainage path is one-half the thickness of the layer.	Time for settlement $= 30 \text{ min} \times \dfrac{60 \times 60}{\frac{1}{2} \times \frac{1}{2}}$ $= 300$ days

Generally, a soil layer is considered free to drain if there are sand layers above and below it. Frequently, clay soils are interfingered with layers of sand. In such a case, the sand layers act as drainage layers, causing settlements to occur faster than for a body of silt or clay without sand layers.

19.5 Acceptable Settlements

Building settlements can be measured as the total settlement of the building, or as the differential settlement between adjacent footings or between the center and corner of the building.

Generally, total settlements can be tolerated without much difficulty as long as they are uniform. If all foundations in a building settle 3 in., the only problems will be to accommodate utilities coming into the building, and the level of the sidewalk and parking areas around the building.

If foundations settle different amounts, for example if one column footing settles 1 in. and a nearby footing settles 2 in., this can create distortion of the building and cracks in walls. This kind of settlement is much more difficult to tolerate. Therefore, differential settlements are of much more importance than total settlements.

Certain kinds of structures, such as large storage tanks, have been known to settle several feet without creating much difficulty in keeping them in operation. Frequently large oil storage tanks are designed and constructed in anticipation of settlements of a foot. The primary requirement is for the tank shell to settle a reasonably similar amount.

In major commercial structures, it is common to limit the allowable differential settlements between adjacent columns to ¼ in. or less. Settlements between adjacent columns might be acceptable for wood frame structures or light steel frame industrial buildings.

19.6 Methods of Accommodating Settlements

Settlements may be reduced by a change in foundation design. This may include larger or deeper foundations. Also, settlements can be reduced if the site is preloaded or "surcharged" prior to construction of the building (see Section 19.7) or if the soil is precompacted (see Chapter 22). If settlements occur quickly, during construction and initial loading, they can be corrected for and will not be a nuisance when the building is taken over by the owner. It is possible to "speed up" settlement by improving drainage of the compressible soils.

Concerning the use of wider foundations, as a rough guide it can be assumed that settlement reduces as the width increases for square footings. For instance, for a column load of 400 kips, assume that the footing size is 8 by 8 ft and the estimated settlement is 1 in. If the footing is changed to 12 by 12 ft, the settlement will be 8 ÷ 12 = ⅔ in. If the footing is changed to 16 by 16 ft, the settlement will be 8 ÷ 16 = ½ in. This guide applies better to silty and clay soils than to sandy soils.

Accommodating settlements is frequently accomplished by changing the structure with regard to its method of framing and wall construction to make it more flexible.

The foundations of light buildings can be designed with long anchor bolts and clip angles on the columns which would permit jacks to be inserted and the columns to be jacked up from time to time to relevel the building frame.

Structural separations may be placed in a building at certain intervals. This would permit some distortion of the building and movement to occur along preestablished "breaks."

Side panels which are clipped to steel columns can accommodate appreciable differential movements without distortion. By contrast, block filler walls, brick walls, poured-in-place concrete, or continuous tilt-up concrete walls are brittle, crack easily, and show distress cracks for even small differential settlements. It should be borne in mind, however, that well-reinforced concrete walls may have considerable girder strength and may tend to bridge over foundations which tend to settle too much. Sometimes building walls are designed to act as girders. This appears to work satisfactorily as long as the required span is not too long. A redistribution of stress does occur in the wall, which is indeterminate.

19.7 Preconsolidation Methods

19.7.1 Surcharge Fills

A building site may be forced to settle prior to construction of the building. Surcharge fills are frequently used for this purpose. Generally, the surcharge fill placed on the site is as heavy as or heavier than the weight of the proposed building. Sufficient time must be available for the settlement to occur prior to construction.

Surcharge fills may be on the order of 4 to 5 ft high for one-story industrial buildings, supermarkets, and school buildings. Surcharge fills

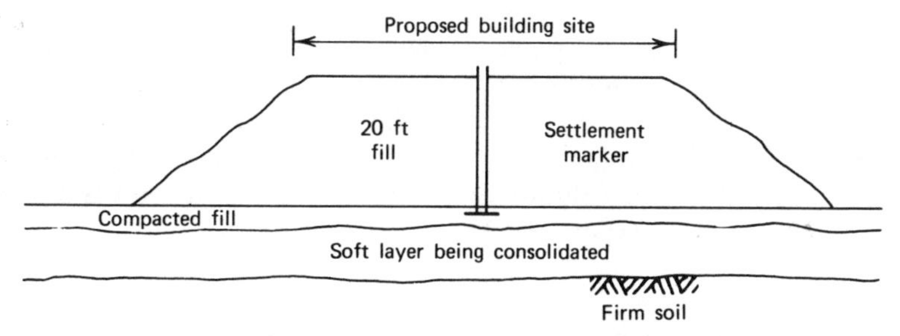

Fig. 19.2 Surcharge fill. Note that the soft layer being consolidated is loaded in the same manner as a laboratory test (see Fig. 12.8). The surcharge process is a "king size" consolidation test. The settlement marker readings give the same answers as can be calculated from laboratory tests.

may be 15 to 20 ft high for relatively heavy reinforced concrete buildings. Occasionally, surcharge fills of 30 to 40 ft are constructed on the sites of proposed heavy storage tanks, power plants, or other extremely heavy structures.

Settlement markers can be placed at the bottom of the surcharge fills to measure the settlement and also the termination of settlement. Figure 19.2 shows a typical surcharge fill.

19.7.2 Compacting or Densifying the Soil

Several methods are available for compacting or densifying the soil. These include (*a*) vibroflotation; (*b*) depressing the water level by use of sumps or well points; (*c*) excavating the soil and replacing it with a compacted fill; and (*d*) inserting sand drains, wick drains, or other devices to permit water to drain out of the soil more rapidly. These methods are described in more detail in Chapter 22.

19.8 Correction

Settlement of foundations has been a problem since antiquity. However, many large and heavy structures which have experienced serious foundation settlements have been stabilized or brought back up to their original positions. The technological and construction capability is available to correct very serious problems of this nature at costs which are reasonable when compared to the replacement value of the structure. Corrective treatments of this type are finally being applied to the Leaning Tower of Pisa.

Commonly used corrective procedures include (*a*) jacking up foundations by injection of soil-cement or other materials under the foundation; (*b*) supporting the column while a new foundation is placed under an existing foundation; (*c*) extending the footing to a lower elevation one portion at a time; (*d*) injecting chemicals or grout into porous soils below a foundation; and (*e*) installing piles, drilled caissons, or other new foundations adjacent to an existing footing, and transferring the column load to these new foundations by means of a "needle" beam. Needle beams may be inserted under a footing, or they may be inserted above the footing and connected directly to the column. A less common method consists of freezing the soil under a foundation and keeping it permanently frozen.

19.9 Settlements of Other Types of Foundations

This chapter has discussed settlements as applied to spread foundations. However, settlements may be a consideration for other types of foundations, such as friction piles, end-bearing piles, groups or clusters of piles, straight shaft piers, drilled and belled caissons, and other deep foundations.

Deep foundations carry their loads down to lower, firmer strata or spread out their loads over a greater depth of soil. These foundations also settle when they are loaded. Usually, the settlements are considerably less than for spread foundations. As a crude guide, if spread foundations were used in a particular area and experienced settlements on the order of 1 to 2 in., it might be expected that pile foundations or other suitable deep foundations would experience settlement of ¼ to ½ in. However, if the foundations should reach end-bearing on hard soil or rock, the settlements would be essentially the elastic shortening of the piles as load was applied.

Assume that a building is supported on friction piles. Friction piles gain their support from the full length of soil through which they are driven, and they do not reach hard end bearing.

If one building column carries a load of 50 tons, it might be supported on a single pile. Another column in the building may support a load of 400 tons and require eight piles. Although each pile is loaded to an equal amount, 50 tons per pile, it should be expected that the eight-pile group will settle more than the single pile. The difference in settlement may not be large; for instance, the single pile might settle ¼ in. and the eight-pile group might settle ½ in. The resulting differential settlement of ¼ in. would not be a problem structurally and would never be detected unless a precise survey was made.

The settlement of individual piles or of pile groups is more difficult to estimate than that for spread foundations. Simplified methods have been developed for rough estimates of settlements. These methods are fairly similar to those used for calculating the settlements of a spread foundation.

Usually, it is assumed that the load on the pile group is spread out, so that the pile group is somewhat equivalent to a spread footing placed at some depth below grade. This concept is illustrated in Fig. 19.3. Having transformed the pile group into an equivalent large spread foundation on soil, the additional soil pressure can be calculated. Using this soil pressure,

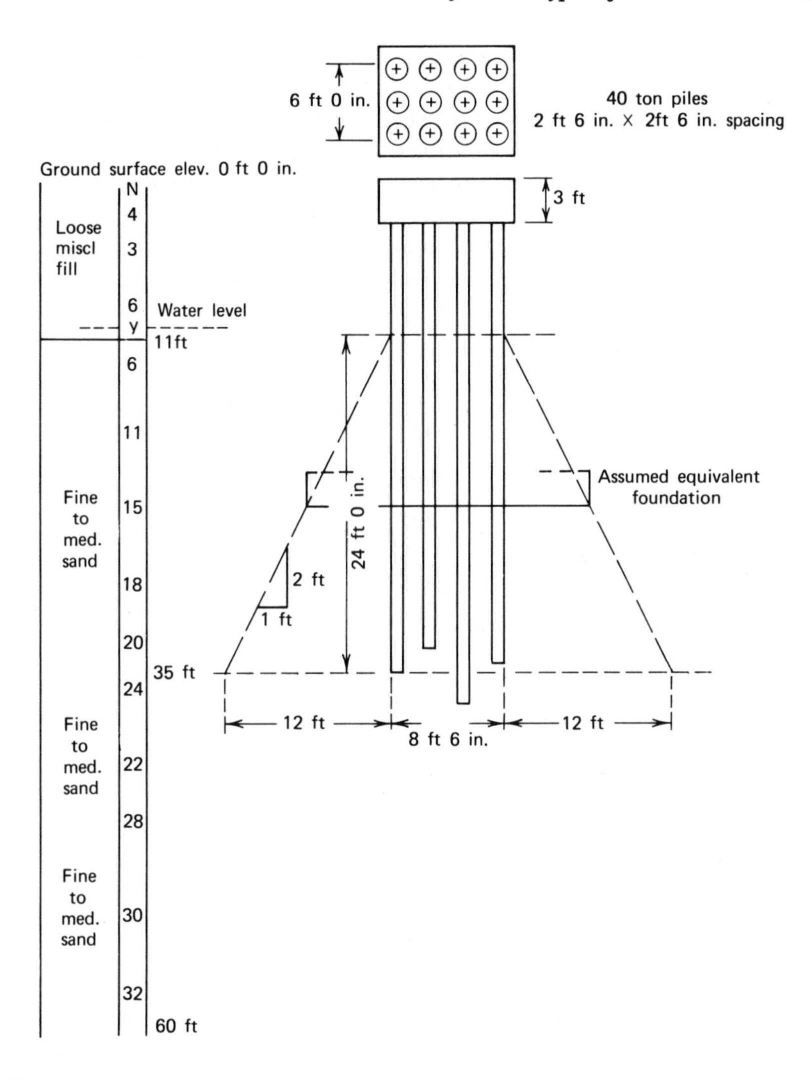

Fig. 19.3 Transformation of a group of friction piles to equivalent spread footing for settlement calculations.

the consolidation of the various soil layers under the footing can be estimated as described previously, and a total settlement can be estimated. By contrast, if the pile group is driven through soft soil, and then to refusal in a deep, very dense sand and gravel, the "equivalent footing" would be deep, as shown in Fig. 19.4.

Usually, settlements of structures supported on piles are relatively small and are not a problem. However, serious difficulties can develop if a portion of a structure is supported on piles while other portions are supported on spread foundations. Figure 19.5 shows a two-story structure which was supported on pile foundations. Adjacent to it, a lighter one-story structure was supported on spread foundations. It was attached to the original two-story building. Large settlements occurred in the one-story building.

As a general rule, if a portion of a structure is supported on piles, other attachments, even though light, also should be supported on pile foundations. Only after very careful study of likely settlements should con-

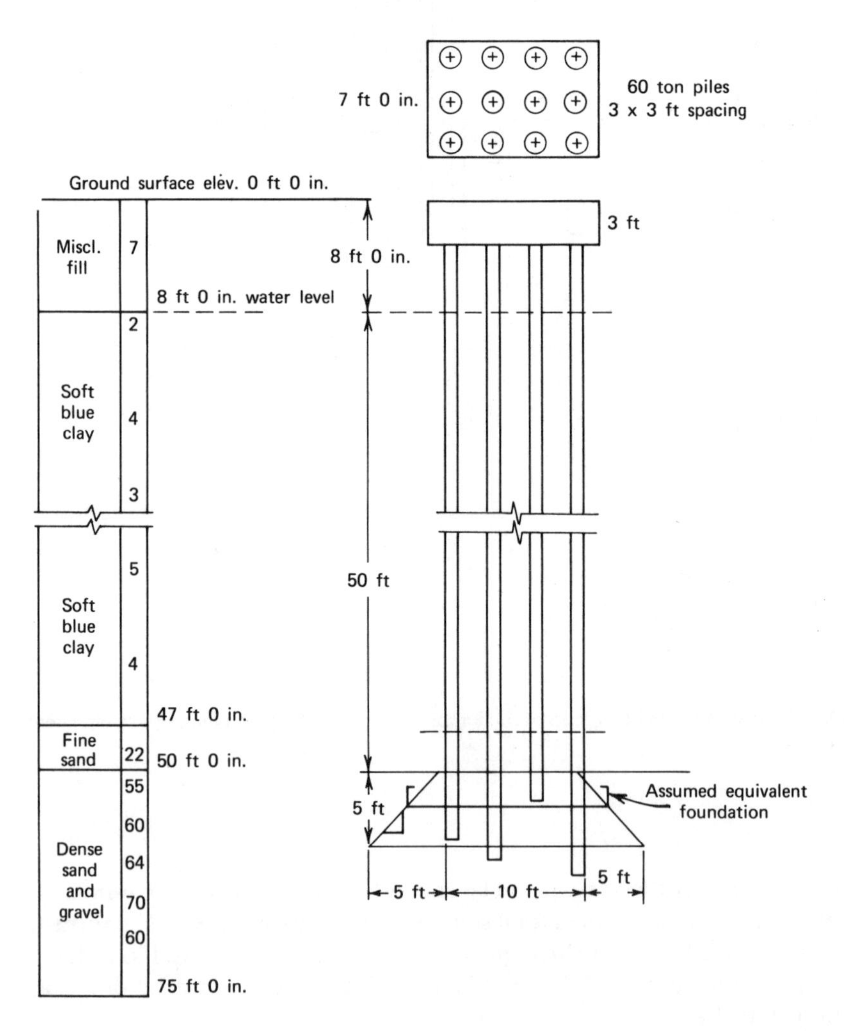

Fig. 19.4 Transformation of group of end bearing piles to equivalent spread footing for settlement calculations.

Fig. 19.5 Differential settlement between pile-supported structure connected to building on spread footings.

sideration be given to using a mixture of foundation types. If the types are mixed, it is best to make a structural separation, with double columns, between the two portions of the structure.

Again, it should be pointed out that this discussion of settlement *does not* tell exactly how to perform a settlement calculation. More information can be obtained from texts, such as Ref. *13*.

20

Pile Foundations

It is the usual practice of the general contractor to sublet the pile foundation work to a subcontractor specializing in pile driving. On small projects, piling may be installed by the general contractor. In this book the discussion of pile foundations is directed primarily toward projects on which a specialty subcontractor furnishes and drives the piles.

20.1 Contract Documents

The contract documents consist of

General conditions.
Form of contract.
Specifications.
Drawings.
Soil conditions (may not be included with contract documents).

Equally binding on the contractor are the local building code and any ordinances such as those regulating the use of hoisting procedures and safety provisions.

While engineers are meticulous in preparing specifications for piling to conform to the requirements of the building code, a contractor should be sure the subcontractor will verify the compatibility of the specifications and the code. Often specifications will restrict the choice of pile types and will contain requirements to fit anticipated soil conditions. These may include acceptable final driving criteria, jetting, or preexcavation of piles which go beyond code provisions.

The selection of the type or types of piles suitable for use on a project is very important. This should be detailed clearly in the specifications. Sometimes specifications can be stretched to include other alternate piles, particularly if money can be saved. A brief summary of pile types is presented in Table 20.1.

The insurance provisions require careful review. The limits of insurance coverage called for may be inadequate for the potential risks. This is particularly true of property damage insurance.

Property damage from pile driving occurs principally from the following: (*a*) rupture or displacement of underground utilities; (*b*) damage to adjacent structures and equipment from vibration; and (*c*) soil displacement. Should utilities be damaged, it is incumbent on the contractor to demonstrate that every reasonable effort was made to discover utilities underlying pile locations. Equipment and services are available to measure the effect of vibration on adjacent structures and equipment (see Chapter 27). Soil displacement is discussed in Section 20.8.

Completed operations insurance coverage should be provided, since claims for damage may not arise until long after the pile work has been completed.

Where the procedures for claims for extra payments require notification of intent to make such claims, strict compliance is important. Changed orders, extra work, or claims based on latent conditions call for the compilation of detailed data, particularly on large, complicated projects where unanticipated work in one area sets off a chain reaction in others. At the earliest moment, competent personnel should be assigned to collect and prepare the backup data for claims. The use of a "critical path method" (or similar system) analysis to demonstrate the overall effect of the extra work on both time and costs should not be overlooked.

A careful review of the soil information accompanying the contract documents will permit certain determinations.

The contract documents may contain a disclaimer as to the soil information. In many cases the soil data are offered to the contractor to use at his or her own risk and with a disclaimer as to their accuracy. Often the contractor is invited to make his or her own soil investigation. The legal aspects affecting any claim for latent or changed conditions are found in Ref. *53*.

Whether the soil investigation for a given site is adequate as to the number of borings, their depth, and the data on soil conditions needs to be determined. Where there appear to be deficiencies, that fact should weigh heavily in the decision as to whether to bid the work. The contractor sometimes elects to make check borings at his or her own expense prior to bidding.

Table 20.1 Pile Types—Advantages and Disadvantages

Pile Type	Advantages	Disadvantages
Timber	Low cost, easy to drive.	Readily broken by overdriving and obstructions. High cut-off waste.
Timber composite	Low cost. Often cheaper than creosoted piles.	Connections problems. Can be difficult to drive.
H-beams	Easy to drive. Able to secure penetration.	Splicing costly. Bottom seating problems. Corrosion a consideration.
Pipe piles	Readily available and economical. Easy to drive. Can inspect bottom. Can be driven in short sections.	Unless splices are welded, can be wet. Friction piles require careful analysis.
Cast-in-place, mandrel driven	Economical. Interior can be inspected. Adaptable to wide length variations. Tapered types beneficial for friction piles.	Heavy specialized equipment needed. Collapse, torn casing, wet pile problems cause numerous rejects.
Cast-in-place, without mandrel	Easy to drive. Readily spliced. Light equipment. Seldom torn or wet.	Limited ability to take hard driving.
Uncased (drilled)	Low cost. Fast. No vibration.	Problems in water-bearing and caving soils. Load tests costly.
Grout injected	No vibration.	Cannot be inspected. Extensive load tests needed.
Enlarged base	Densification of loose granular soils. High bearing capacity.	Cannot be inspected. Casing needed through compressible soils.
Precast	Permanence. High capacity. Strong in bending.	High cut-off waste. Hard to handle. Jetting needed to assist driving. Easily broken.
Prestressed	Permanence. Resists corrosion. High capacity. Strong in bending. Can take hard driving.	Vibration during hard driving. Careful handling. Heavy equipment. Needs jetting. Costly cut-off waste.

It may not be feasible to drive piles by usual methods to specified depths as based on a review of the soil borings. They may show that special methods, such as jetting, will be needed.

20.2 Bases of Submission of Quotations for Piling

Various forms for requesting quotations for piling work are employed. The three most frequently used are as follows:

Quotations are based on a fixed number of piles of an average length. The price submitted is known as the "principal sum." Unit prices are submitted for adjustment of the principal sum for the addition or omission of an average length pile. A price per lineal foot (l.f.) is allowed for piles longer than the average length and separately a credit per l.f. for omitted lengths of shorter piles. In some cases, contractors insert their own unit prices; in other cases, the engineers determine what they shall be.

Bids may be requested for a lump sum for mobilization and demobilization of equipment, plus a unit price per l.f. based on an aggregate footage for the work.

A lump sum quotation is obtained for a project irrespective of the actual lengths of piles driven. The only adjustments to be made are for added or omitted piles, for which unit prices are submitted.

Of these forms of bidding, the use of a principal sum with adjustments may produce the lowest price because the piling contractor assumes the least risk from length variations.

20.3 Estimating the Bearing Capacity of Piles

The capacities of piles are estimated by two basic methods. The derivation, application, and limitations of static analysis of capacity and of pile driving formulas are covered in detail in engineering literature (see Refs. *54* and *55*). Therefore, the following is only a brief outline of the basic considerations.

In the static analysis method, the load-bearing capacity of a pile is estimated from an evaluation of the soils that provide support for the pile. The soil provides end-bearing support and side friction support. Both can be estimated.

The soil profile must be developed from borings, pits, or other explorations. The strength of the various layers in the soil profile can be measured by laboratory tests on samples of the soil or by resistance to penetration. Penetration resistance is usually measured by using a soil sampler, cone, or other similar device driven or pushed into the soil.

For each layer of soil, the frictional strength is calculated based on the test data. The strength may be adjusted for the side pressure around the pile developed when the pile is driven and shoves the soil sideways. The frictional strength multiplied by the surface area of the pile gives the support in each layer. The support values for the several layers are added up. Also, the bearing capacity of the soil at the proposed pile tip depth is estimated using common bearing value formulas, but probably modified for the great depth of the tip. All of the capacity values are added up to obtain the ultimate capacity of the pile. This value usually is reduced to two-thirds or one-half to obtain a "design capacity."

The pile driving formula was developed over a period of many years by comparing driving resistance to results of load tests. The formulas usually consider the weight and fall of the hammer and the number of blows to drive the pile an inch. Further accuracy can be obtained by considering efficiency of driving, the weight of the pile, and other factors. The simpler formulas, such as the *Engineering News* formula, are most commonly used.

As an example of the use of the *Engineering News* formula, take the following data:

$$\begin{aligned}\text{Hammer weight} &= 5000\text{ lb}\\ \text{Hammer drop} &= 3\text{ ft}\\ \text{Rate of driving} &= \tfrac{1}{4}\text{ in. for each blow}\end{aligned}$$

$$\begin{aligned}\text{Capacity} &= \frac{\text{weight} \times \text{drop} \times 2}{\text{rate of driving} + 0.1}\\ &= \frac{5000 \times 3 \times 2}{\tfrac{1}{4} + 0.1}\\ &= \frac{30{,}000}{0.35} = 86{,}000\text{ lb}\\ &\qquad = 43\text{ tons (factor of safety} = \text{approximately 2)}\end{aligned}$$

20.4 Test Piles and Pile Load Tests

In the design of pile foundations, presumptive pile lengths and bearing capacities are determined by engineers prior to construction. These estimates sometimes are verified by driving test piles and conducting load tests on some of the test piles. The test piles also provide the basis for ordering one-piece piling, such as timber or H-beam piles.

The need to drive test piles for ordering materials is independent of the presumptive bearing loads. Virtually all building codes exempt lightly loaded piles (up to and, in most cases, including 40-ton working loads) from pile load tests. Verification of bearing capacity by pile driving formula is accepted. The *Engineering News* formula, sometimes with modifications, is generally specified. Load tests are called for by engineers under special circumstances. For example, piles may be driven to a predetermined depth and driving may be suspended before the indicated final driving resistance has been attained. Load tests may then be conducted to satisfy both the engineers and the building department.

Where piles are designed for working loads in excess of 40 tons, a test pile and load test program may be carried out in advance of production pile driving. Otherwise, detailed soil tests and static analysis methods are required to select pile penetration to develop the design load capacity.

The number of test piles to be driven depends on the code requirements, the size of the site, and soil conditions. Test piles are usually driven in pairs and should always be located close to test borings. At least one pair of test piles should be driven in the area of poorest soil conditions. Pile load tests are performed on a selected number of the test piles. Test procedures usually conform to an accepted standard, such as ASTM D 1143, *35*.

Pile load tests can be simple and straightforward or they can become field research projects involving special devices to measure the movement of the tip of the pile as well as the settlement of the top. Field personnel should be made aware of the importance of meticulous care in record keeping, including temperature readings affecting the gauges. On important work, separate settlement readings with surveying instruments are insurance against misreading or malfunction of the gauges. Field personnel in their well-intentioned zeal sometimes feel an urge to overdrive the test pile, "just to make sure." Since the piles have to be driven to the same final depth and resistance as the tested pile, this practice should be discouraged as it can backfire to everyone's disadvantage. Much has been written in engineering literature on the subject of pile load tests and the evaluations of the results (see Refs. *54* and *55*).

Several criteria must be met for a load test to be considered satisfactory. Loading is increased in increments to a total of 200% of the working load. Unloading in increments is usually required. Generally the maximum permissible settlement under 200% loading is 0.01 in./ton of applied load (gross settlement). Many codes, however, limit the gross settlement to 1 in. or apply some other limiting factor. In many codes the gross settlement is not limited, but the maximum net settlement (after unloading) is specified. Nearly all codes require that there be no settlement under full load for periods ranging from 6 to 48 hr.

Experience has shown that the requirements of 1-in. net and 0.01 in./ ton of load for gross settlement usually have resulted in satisfactory pile foundations. More restrictive provisions generally are not required, unless there are severe limits on settlement or some other performance criteria.

20.5 Verification of Estimated Pile Lengths and Bearing Values

In preparing estimates for pile foundations, contractors must review the soil data, the driving specifications, and the test pile and pile load test program to arrive at independent conclusions as to the average length of the piles compared with the lengths on which the bid is based, difficulties anticipated in fulfilling the driving specifications, and any risk of failure of the pile load tests.

In considering the probable pile lengths, some of the factors are the compaction of the soil where there are a large number of piles in groups, the presence of mica or silt which act as lubricants in granular soils, and the relief of overburden pressure due to removing the soil from deep excavations.

Driving requirements for purely friction piles seldom involve more than meeting a fixed length or a final driving resistance, or both. Where a specified bearing stratum must be penetrated, difficulties can arise. The top of the stratum may be defined clearly at each boring location. Between borings, however, the assumption of straight-line interpolation of its surface may result in considerable overdriving to reach the assumed tip elevation. When high-capacity piles must be driven to rock and rock exists at some elevation higher than the interpolated elevations, efforts to demonstrate that rock has been reached can involve overdriving and possible damage to the pile and to equipment. Also, there may be the time and cost of additional borings to verify the rock surface.

Should it appear that the bearing capacities are critically high for the soil conditions, the load tests may fail. The consequences must be considered, as well as alternatives available to the engineers. Not only the estimate of cost but also the selection of equipment must take such possible problems into account.

Frequently, soil conditions at a site turn out to be different than what was expected. This may be due to meandering streams with varying deposits of sand and silt, an erratic surface of rock or hard material overlaid by more recent softer material, or the presence of sinkholes in underlying limestone rock.

As a result, driving of piles is very confusing. The pile lengths vary considerably from one pile to the next. The driving record also may vary considerably from one pile to the next. How can the contractor decide on the proper pile length and the capacity of the pile?

These questions cause delays. Also, there are additional costs for making some piles longer or possibly even abandoning some piles.

In such cases, borings drilled at a site may have been located in such a way as to encounter firmer soils or to hit the bedrock or hard material at the higher elevations. In between, the rock is deeper or the soils are softer. Almost always, the only reasonable way to solve the problem is to obtain additional information. Additional borings should be drilled. Generally a considerable number of additional borings are drilled to make sure that the conditions are clearly identified.

Where additional pile lengths are required, who should pay? If the true soil conditions had been known in the beginning, the longer pile lengths would have been bid originally and would have been paid for by contract. The soil conditions did not change—they were a part of the site and belong to the owner. Therefore, in almost all cases, the cost of additional pile lengths is paid by the owner.

20.6 Pile Hammers and Equipment

Pile driving hammers in general use for foundation piling and other construction applications include single-acting, differential, diesel, and vibratory hammers. They range in size from those having energy ratings below 8000 ft-lb per blow to hammers rated at nearly 40,000 ft-lb per blow. There are lighter hammers available for driving timber sheeting and other uses and heavier hammers for special uses, such as marine construction.

Hammers with energy ratings less than 15,000 ft-lb are employed mainly for driving steel sheet piling and soldier beams in short lengths (under 25 ft) and for easy driving conditions. Lightly loaded timber piles under 25-ton capacity have often been successfully driven with hammers rated at 12,000 ft-lb of both single-acting and double-acting types.

The minimum energy rating permitted by most building codes for bearing piles is 15,000 ft-lb. For many years, the hammers available were limited to maximum ratings of 15,000 ft-lb. (A few heavier hammers were available.) During that period, the bearing loads on piles rarely exceeded 60 tons. As bearing loads were increased, larger hammers were developed, and concurrently the differential hammer was introduced. Engineering practice and the more modern codes call for the use of higher energy hammers for heavier pile loads. For example, the New York City code accepts 15,000-ft-lb hammers for piles up to 60 tons, and calls for 19,000-ft-lb for 70- to 90-ton piles. Over 90-ton capacity, piles must be driven with 24,000-ft-lb hammers.

Where the soil offers considerable resistance to driving, a high-energy differential hammer, which hits twice as many blows per minute as a single acting hammer, will give greater production. But where the final driving resistance is high, the heavier hammers may cause damage to the tops and tips of the piles. This is an important consideration in selecting hammers.

Diesel hammers, as the name implies, use fuel oil pumped into the combustion chamber which, mixed with high-temperature air under pressure, ignites and creates the energy for each blow of the hammer. The need for boilers or compressors is eliminated. The range of energies which diesel hammers provide makes them suitable for most types of piles and pile capacities.

Single-acting diesel hammers operate somewhat slower than single-acting steam hammers. Diesel hammers have to be started with a trip; thereafter their operation is automatic. In penetrating soils with little or no resistance, the trip may have to be used until some resistance to driving is encountered.

Vibratory pile hammers have been developed in relatively recent times. The hammers get their name from the high frequency of the vertical strokes developed by the driving mechanism. They are useful in driving and extraction of steel sheeting and soldier beams. They have also been used to a limited extent for bearing piles of low to medium load ranges. Claims have been made that vibratory hammers can drive timber and pipe piles successfully through soils containing obstructions and through which conventionally driven piles could not penetrate. While there is some evidence to support the claim, more proof is needed and may come in time.

Resonant (so-called sonic) drivers impart vertical high-frequency vibrations to the pile from a self-contained "driver" mounted on the front of the pile driver leaders. The vertical vibration of the pile reduces frictional resistance on the side of the pile so that nearly all the energy is transmitted to the pile tip. The energy is sufficient to break or dislodge boulders, penetrate obstructions, and pulverize ledge rock.

Most of the vibration caused by impact pile driving is eliminated. The noise generated by resonant driving is less objectionable than that generated by impact driving. Under most soil conditions, the time required to sink a pile is much less than for impact driving. Also, the piles are straighter. These units are especially effective in sandy soil and less effective in clay soil.

Resonant driving has been successful from an engineering and cost standpoint. However, further research is needed for the control of repair costs of the hammers to make this promising system more competitive.

Pile extractors are made by the major hammer manufacturers. Extractors may be double acting or single acting and rigged to drive upward. Vibratory extractors drive down or up and depend on vibration to break the friction and reduce the pull required to extract the pile. The vibratory driver-extractor appears to be the best for pulling steel sheet piling. For other piles, extraction depends on the pulling power of the crane aided by the friction-breaking action of the extractor. The effective line pull of the crane is limited to the tension that the extractor can withstand and the grip on the pile. There is always the risk that the grip on the pile may let go.

Precautions must be taken to provide some check to the boom against backward rotation if a break should occur.

Driving piles down an inch or two will often temporarily break the soil friction. In granular soils, a water jet may be all that is needed to loosen a pile.

The present generation of pile drivers are caterpillar-mounted cranes. For special projects, whirleys mounted on bed sills which in turn move on rollers permit a much greater radius for driving than crane drivers. Pile hammers are handled with a maximum of four parts of wire line reaved over sheaves at the top of the leaders. The leaders may be attached to the boom by a connection at the top, in which case they rest on blocks on the ground during driving, or they may be connected so that the leaders extend well above the end of the crane boom. A frame or spotter can be adjusted in length so that piles may be driven on a batter either in or out. Special rigging is necessary to position the leaders to drive batter piles sideways.

20.7 Pile Driving

The process of pile driving is of interest primarily because so many things can go wrong. For instance, human-made surface fills often contain broken concrete, rip rap, timber, scrap iron, and the like. Such obstructions can break timber piles, tear the casing of cast-in-place piling, and cause piles to drift from location. In natural deposits, obstructions consist of boulders, rock fragments, or gravel measuring 6 in. or larger. Excavations are often made at pile locations to insure good pile production through shallow obstructed fill. Holes can be dug with a backhoe. After discarding the obstructions, the hole can be backfilled with the remaining soil, and the pile can then be driven. Sometimes a mobile auger drill is used ahead of the driver to test for obstructions. Where there are so many piles that the pile caps almost form a mat, consideration should be given to excavating and backfilling the entire area.

Spuds are sometimes used where obstructions are located below the water level or are too deep for removal by inexpensive excavation. If the obstructions lie in loose or soft soil, they may be shoved sideways by driving some form of spud. The spud may be a heavy H-beam with the tip reinforced or a heavy wall closed-end pipe. It is often necessary to drive the spud more than once in a group of piles, since pushing the obstruction from one pile location may drive it to another. Where obstructions cannot be dislodged, it may be necessary to use the spud to locate unobstructed areas within the foundation area, drive the piles where they can penetrate, and redesign the pile cap for the piles as driven.

Predrilling is required frequently to get through firm soil layers or obstructions. In steel mill areas there are often slag fills up to 30 ft deep.

Such fills contain "skulls," hardened, rounded lumps of iron, or slag from the bottom of the ladles. Where the skulls cannot be spudded out of the way, it has been found necessary to drill auger holes at pile locations, and where skulls are encountered to blow them with dynamite.

The process of "casing off" may be used where sharp fragments will tear the casing of cast-in-place piles. An oversized casing is first driven and cleaned, and the pile driven through the casing which protects the shell of the pile after which the casing is withdrawn. Under some conditions, the casing can be driven with a loose-fitting pan on the bottom, which must be driven off before the pile is driven through the casing.

20.8 Ground Heave and Displacement

Most types of piles are "displacement" piles; that is, they displace a volume of soil about equal to the volume of the pile. H-beam piles and open-end piles are considered "nondisplacement" piles. Actually, H-beam piles do displace some soil because a soil plug forms between the flanges and moves down with the pile as it is driven.

In granular soils, the displaced volume results in compaction of the soil surrounding the pile. Where granular soils are too dense for further compaction, the piles may have reached their bearing. If not, jets will be required to wash out the soil to permit further penetration.

In clay soils, driving displacement piles creates high stresses in the soil. Such soil is relatively incompressible during driving and can be relieved only by movement. The soil moves upward and also moves laterally away from the pile. The soil movements take time; thus the continuous driving of piles causes buildup of stress. On a level construction site, where the general excavation is shallow, the ground surface will heave upward by about the total volume of all piles driven. With 10 or more piles in a group, with pile caps which are close together, and with long piles displacing 1½ yd^3 or more, the whole area will heave vertically. The effect may be noticeable as it tapers off beyond the horizontal limits of the excavation. Utilities under adjacent streets may be displaced and heaved, and adjacent buildings may be affected. The driving sequence can have some influence on the heaving of piles. In a large group of piles, the sequence of driving should start in the center and work out toward the perimeter of the pile group. This permits more relief of soil stresses and reduces movement of driven piles. Should the site be located near the top of a deep cut or river bank, both vertical and horizontal movements may occur. Under extreme conditions, the whole soil mass may be set in motion so that pile groups already driven may move horizontally several inches or feet.

When shell-type piles are driven, steps must be taken to verify the integrity of the driven piles. The usual procedure is to place a long piece of 3-in. pipe in the first pile driven in each group so as to establish the elevation of the tip and also the top of the pile. Where the tops of the piles heave but the tips do not, and where no separation of the casing at any joint is observed, the piles may be concreted after all motion within the heave range has ceased. Any lateral motion must also cease. Added piles may be required to correct for eccentricity. Should the pipe telltale reveal upward movement of the pile tips, redriving (see Section 20.10) becomes necessary. To avoid inviting horizontal movements, it has been found advantageous to drive all piles from one level and make excavations for pits and partial basements after the pile work is finished. A cutting-off tool can be used to cut pile casings below grade.

20.9 Preexcavation and Jetting

Lateral displacement within deep beds of cohesive soils (as described in Section 20.8) can be avoided by removing a quantity of soil at each pile location which represents all or part of the volume of the pile to be driven. The procedure is called preexcavation, preboring, or predrilling. The first attempt to do this was with a single-tube preexcavator. It consists of a length of 14- to 16-in. pipe open at the bottom and closed at the top, with a connection at the top to admit steam or compressed air. The preexcavator is driven into the soil at each pile location and withdrawn. The plug of soil is expelled from the tube by steam or air pressure. It should not be used in soft soils, which flow back into the hole when the tube is pulled. It can be used in firmer soils but is limited to depths of about 30 ft maximum. In suitable soils, preexcavation may be accomplished with an earth auger operated by mobile drilling rigs. The augered hole must be able to stand open until the pile is driven. The success of preexcavation depends on close control of the actual volume of soil removed. Ideally, the preexcavation should conform to the outside dimensions of the pile and remove 90 to 95% of the pile volume above the bearing stratum. Thus for piles over 50 ft long, dry tube preexcavation to a depth of approximately 30 ft helps reduce heave, but does nothing to relieve the stress and resultant soil movement below the 30-ft level.

Occasionally soil conditions are such that there is a reduction in pile driving resistance after pile driving is stopped. Where a pile has been driven initially to a specified final resistance of, say, eight blows to the last inch and after a pause of 15 min or more driving is resumed, the resistance may drop to three or four blows per inch. The previous final resistance has been partially lost. It may well take another foot or more of penetration to regain the specified resistance. Such a phenomenon has been called "relaxation of pile driving resistance."

Experience to date points to some shales and dense fine silt or sand as formations in which piles may behave in this manner. Loss of resistance has occurred in friable shales, particularly when interbedded with soft seams, and in saturated fine silts and sands where the N value has been 50 or higher.

One proposed explanation is that pore water, unable to move rapidly through dense fine soils, resists a large proportion of the hammer energy under repeated blows. The pore water pressure returns to normal during the pause in driving, permitting easier penetration when driving is resumed.

There is no single cure for this dilemma. One method has been to drive the piles initially to a resistance 50% greater than the specified resistance, and then, by redriving, determine whether the specified resistance has been retained. This system requires that every pile be redriven to prove its capacity. Also, additional pile load tests may be required (see Ref. *56*).

To provide a reliable, controlled method of preexcavation for piles of any length, especially in soft, cohesive soils, "wet-rotary" preexcavation was developed. In this system, a hole is drilled to the desired depth by rotary drilling methods. The hole can be straight or "shaped" by reamers attached at selected positions on the drill stem. Drilling mud is pumped through the drill stem and recirculated to a sump. The drilled hole is left full of slurry to prevent squeezing or sloughing of the walls. Thereafter the pile is driven into the hole, displacing the slurry and developing intimate contact with the soil for the preexcavated lengths.

Jetting of piles, in a sense, is also preexcavation. Water jets are used to create a hole into which piles are driven, or to relieve piles of friction as they are being driven; this method is about as old as the art of pile driving. Although it is feasible to jet holes through sandy soils, it accomplishes little in clay soils. In sands, a typical setup is a 3-in. jet pipe with a nozzle on the end to jet or wash a hole into which the pile may be lowered and then driven. The presence of gravel in a formation often limits the effectiveness of jetting because the gravel tends to accumulate at the bottom of the jetted hole. Jetting alongside of the pile is probably more common. With a single jet, it is difficult to keep the pile on location. Therefore, piles are more often jetted with twin jets, one on each side of the pile, for greater effectiveness and better control.

20.10 Redriving and Tapping of Piles

The heaving of cohesive soils during driving frequently causes adjacent piles to lift with the soil. This causes the tips to lift off the bottom. Possibly the piles will stretch and separate if the tips are anchored in a strong bearing stratum.

One-piece piles such as timber piles, precast concrete, H-beams, and pipe obviously must heave as a unit. They can be "tapped" back to the original tip elevation and driving resistance by swinging the pile hammer back onto them.

In the case of shell piles, a telltale pipe inserted in the shell down to the tip will demonstrate the relative movement of the top of the shell and the tip of the pile. Evidence of movement of the tip calls for redriving. The mandrel must be inserted in the pile casing and the pile redriven to the same final resistance with the same energy hammer. Distortion of the casing, dog-legs, or bends in the casing can prevent the reentry of the mandrel. In such cases, redriving tools (consisting of a heavy walled pipe smaller than the mandrel, and fitted at the bottom to conform to the tip of the mandrel) are often used to redrive distorted piles. Mandrel-driven cast-in-place piles, with corrugated thin shells, can stretch to some degree without separating before concrete is placed.

Where the bearing stratum is primarily clay soil, it may heave, carrying the entire pile with it. It has been demonstrated many times that the bearing capacity is not necessarily affected. This, however, has to be proven in each instance. The sequence of driving piles in a large group can influence the amount of ground heaving. It is best to start driving at the center of the group, and work outward.

More than one redriving or tapping may be needed to make sure that all piles in a group are seated properly. For cast-in-place piles, concrete should not be poured in piles until all heave has ceased and all piles are seated. Concreted piles may be redriven, but this involves some risk and is not good practice because of potential damage to the concrete.

20.11 Negative Friction of Piles

The preparation of a construction site may include raising the general grade by placing fill. When the fill is placed over a clay formation (including peat), the weight of the fill causes consolidation of the clay bed. The entire site settles, which may occur over an extended period of time. Piles that are driven through the fill, the clay, and then into a bearing stratum derive most of their support from the bearing stratum. Some temporary support is contributed by friction in the fill and clay. As site settlement proceeds, the fill and the clay pull down on the piles and transmit load onto the piles. The piles may settle under this load (see Ref. *54*). The negative friction or "drag-down" is in addition to the building loads. The amount of drag-down on the piles can be estimated as a function of the average strength of the fill and the clay. A technical soils analysis is essential for designing pile foundations under these circumstances.

20.12 Uplift Resistance

The ultimate capacity of a pile in tension is generally assumed to be measured by the shearing resistance of the soil surrounding it or the amount of adhesion between the pile surface and the soil, whichever is lower. There is evidence, however, to show that in some soils, moderately firm clays, for example, failure may occur below these values. Consequently, it is often advisable to verify uplift capacities by field testing. The usual procedure is to drive two reaction piles flanking the test pile. Then with a jacking frame, lift up on the test pile until failure occurs. By reducing the loading to zero at three or four points in the loading, the movement of the pile relative to the soil and elastic lengthening of the pile can be detected. The results are plotted in the same way as for a downward load test. From these data, and using an appropriate factor of safety, safe uplift load can be established.

Generally, for friction piles the uplift load is the range of 50 to 75% of the downward bearing capacity. For end-bearing piles, the uplift capacity may be small.

20.13 Lateral Resistance of Piles

Several methods have been developed for estimating the lateral loads which can be put on piles. Generally, they consider the load which can be placed on a single pile. Four commonly used approaches are as follows:

Pole formula. Several pole formulas have been developed, considering basically how deep a flagpole must be embedded to keep it from overturning. The pressure on the faces of the pole are indicated in Fig. 16.8. It is necessary to know the soil's allowable lateral bearing value to make the calculation. Pole formulas are given in some building codes, such as the uniform building code. A typical formula is given in Fig. 16.8. Allowable lateral bearing values are given in some codes, such as the uniform building code, or can be obtained from soils engineers.

Point of fixity. In this method, it is assumed that the pile is fixed at some point in the earth and is a free cantilever above that point. Points of fixity are arbitrarily assumed. Commonly used depths of fixity are 5 ft below ground surface for firm or compact soil and 10 ft below ground surface for soft or loose soil.

Elastic analysis. In this method, the soil elastic deformation is calculated. The result is a diagram of lateral deflection of the head of the pile for various assumed lateral loads. Since lateral deflection cannot be too great without damage to the structure, the allowable lateral design load on the pile is established. Soil testing and analysis are required for this method.

Load test. A pole or pile can be driven and tested by applying a lateral load at the head of the pile. Usually, a lateral deflecting of ½ in. is the maximum allowable, with residual deflection after removal of load of ¼ in. or less. Also, the design load is usually 50%, or some other percentage, of the test lateral load.

Additional information on design can be obtained from Ref. 57.

21

Site Grading

21.1 Site Preparation and Prerolling

The first step in site grading is drainage. If the site is wet, immediate draining is cheaper than working in the mud. If the site is dry, protective ditches, berms, or other facilities may be needed for future rains or possible flooding. The second step is to prepare the existing ground surface. In hillside areas, this may involve removal of trees, brush, grass and weeds, and old mud flows or landslide debris. In lower areas, organic soils, trash and dumped debris, expansive or unsuitable topsoil, or topsoil containing roots and organic material must be removed. Considerable expense may be involved in removing these unsatisfactory materials and finding a suitable place to dispose of them. Therefore, the contractor should have a good knowledge of the site conditions and of the soil below original ground surface.

Basic data, which should be available and which the contractor should obtain and review, include

1. A profile of the subsurface soil layers.
2. Unfavorable bedding or fracture planes in bedrock formations.
3. A general idea of the firmness and stability of the underlying soils or rocks.
4. The groundwater level, and whether soils to be used may require drying.
5. Soils that are considered unsuitable and require removal.

Specifications are generally vague regarding the amount of organic material which can be contained in soil. Almost all soils contain at least some organic material. In most cases, soils containing a small amount of organic material are acceptable in compacted fills. It becomes objectionable if there are large pieces, roots, or clumps of organic material or vegetation. If it is small and well distributed, it is likely that 2 to 4% of organic material would be accepted in fill soils for most structural fills. In some cases, it may be sufficient to scrape the grass and brush off the surface. The topsoil and roots can mix with the underlying soil and become "lost." By contrast, however, some high-quality structural fills under pavements or major structures may require essentially no organic content. This may require stripping off vegetation plus 4 to 6 in. of topsoil and removing it from the site.

If a soil contains some roots, it may be economical to spread the cut soil on surfaces to be filled, and then rake the fill soils to remove the larger roots. A heavy-duty rake is required for this purpose, but may effect a substantial savings in making a particular soil acceptable.

In some cases, chipping machines are used to chop up the sparse brush on a site and spread it uniformly through the borrow soil. So long as the chips are small and well distributed through the fill, this may be an acceptable fill.

The big question is, how much organic material can reasonably be left in fill soil? If a fill is required to be compacted to 90% relative compaction, it would be possible to overcompact the soil to make up for the space occupied by the roots. Assume that the organic material is 3% by weight of the total soil. The roots have a lower density than soil; therefore, 3% by weight would be perhaps 5% by volume of the soil. If the soil were compacted to 95% relative compaction, the roots could decay and disappear, with a resulting drip in relative compaction to 90%. Overcompacting may be cheaper than attempting to get all of the roots and organic material out of the soil. However, it will be necessary for tests to ensure that this procedure will work on the particular soil involved, and it will also be necessary to get approval of the engineer or owner to use this procedure. Ignition tests may be required to evaluate the quantity of organic matter in the soil.

Where large amounts of organic material occur in a soil and it is not economical to remove a substantial part of the organic material, this soil should be considered for landscaping purposes or for areas in which a high-quality structural fill may not be necessary. Such a fill material, after being well compacted, may have satisfactory stability against sliding or erosion, even though it is not satisfactory for support of structures. If adequate information is not available, one or two days of digging test pits with a backhoe or auger may be a good investment to determine the quantity of soils to be wasted and the amount of usable soil.

Prerolling or proofrolling is a common practice to test a site. Before starting to place fill on a site, or before placing a base course or a floor slab on the soil subgrade, specifications frequently require proofrolling. A loaded dump truck or a compaction roller is used to roll back and forth over the site, followed by an inspector who watches for any soft spots. Frequently soft pipeline backfills or "weaving" plastic soils are found. These soils usually are dug out and replaced with dry sandy soil. Proofrolling is commonly done on completion of airport runways to confirm adequate compaction of the subgrade. A 50-ton or heavier rubber-tired (inflation 90–120 psi) roller is commonly employed.

21.2 Cuts and Fills

Generally, a grading plan is prepared by the engineer to outline the areas of cuts and fills, the depths of cutting and filling, and other details required for the completed rough grading. The grading plan usually is based on balancing the cut and fill. After deducting the volume of any soil which must be wasted, the amount of cut is made to balance the volume of required fill. However, the volume of soil excavated may not exactly equal the volume of compacted fill. One cubic yard of soil excavated may equal only $^{8}/_{10}$ yd of fill in compacted embankment. This is described as shrinkage. Shrinkage can consist of several factors such as:

When soil is compacted, it may have a higher density than the natural soil. Therefore, more grains of soil are contained in each cubic yard of fill embankment. This results in reduced volume of the embankment soil.

Calculation of fill volume is based on compacting the soil to the required percent of compaction. If the specifications require that the fill be compacted to 90% (of the specified method of compaction testing), but the contractor actually compacts the soil to a density of 93%, there is an additional loss of about 3% of fill volume.

Some soil may be lost by spilling during hauling from the cut area to the fill area.

Fill embankments may be constructed somewhat larger than shown on the plans. For instance, a roadway or canal embankment may be specified as 10 ft wide. However, in construction the embankment turns out to be 10 to 12 ft in width. Unless the contractor goes back and carefully trims the embankment to size, there is additional loss of fill material.

Shrinkage values in many cases are in the range of 10 to 25%, but sometimes are over 30%.

The grading plan usually specifies the required slopes for cut areas and for fill areas. Some typical slope angles are as follows:

Type of Construction	Cuts in Rock	Cuts in Soil	Compacted Fills
Housing subdivisions, uniform building code	2:1	2:1	2:1
Dams			3:1
Buttress fills			2:1
Canal slopes, medium height	1:1	1½:1	1½:1
Canal slopes, under 6 ft high	½:1	1:1	1:1

Note. 2:1 means 2 horizontal to 1 vertical.

Although many building departments list acceptable slope angles for cuts and fills in the building codes, all codes permit variances. The variances depend on careful analysis by a foundation engineer to convince the building department that a slope angle is safe and reasonable for a particular site and type of soil.

During construction of fills in hillside areas, it is considered good practice to cut steps into the hillside to bond the fill to the hillside material. This is called benching. The height of risers for each bench may be equal to the height of each lift of fill, or may be higher, say 2 or 3 ft high, depending on which is most convenient to the grading operation.

In the transition from cut to fill, trouble frequently occurs. The fill area tends to settle, while the cut area does not. This unequal movement can crack house foundations and break water mains. Where this condition occurs under house sites, a deeper foundation or shifting the house location should be considered. Pipelines should be made more flexible in these locations by using short lengths of pipe or pipe designed to bend.

Boulders may be encountered in the excavation. Disposal of boulders can be a serious problem. Also, disposal results in a loss or shrinkage of available fill material. Sometimes boulders can be used in the embankment fill. This is described in more detail in Chapter 22, Section 22.12.

Site grading also may be required on flat properties to be used for industrial or commercial construction. Specifications frequently require "proofrolling" of the site prior to placing fill. This is especially true for industrial buildings, which may have heavy floor slabs or floors at dock height which will require placing 3 to 4 ft of fill (see Section 21.1).

21.3 Site Drainage

During grading, construction sites are particularly vulnerable to rainy weather. Therefore, good site drainage should be developed as part of the grading plan developed by the contractor. The contractor's grading plan must use the project grading plan as its end result, but the contractor should plan the progress of the various cuts and fills so that he or she can obtain the most efficient use of the equipment. In addition, the grading plan must be such that the site is not in a vulnerable condition during a rain which would cause the site to become swamped.

In many parts of the country, there is a rainy season. In some cases, it is almost impossible to run an efficient grading operation during rainy seasons. Therefore, the grading contractor may elect to close down grading operations during this season. In some cases, city codes will force closing down grading operations in hillside residential areas during the rainy season.

When a site is closed down for the rainy season, more attention must be given to good drainage, and to finishing off the surfaces of existing fills and cutting areas to reduce the infiltration of rainwater and to reduce erosion (see Chapter 22, Section 22.11).

22

Compaction of Soil

The use of compacted fills and the techniques for construction of compacted fills have developed rapidly over the past 40 years. The development of larger and more powerful excavation and hauling equipment has made grading operations more economical. This has led to the development of many kinds of compaction machines, and the use of compacted soils for many purposes.

This chapter describes the construction of compacted fills *as they are placed* in major earthmoving operations. Compaction of backfill around structures and over pipelines is described in Chapter 17. Compaction of preexisting fills or low-density natural soils "in place" is described later in this chapter.

22.1 Purposes

The major purpose of compacting soil is to make a change in the mechanical characteristics of the soil. These changes include the following:

1. An increase in the strength of the soil—higher bearing value.
2. A decrease in the compressibility (or settlement) of the soil under load.
3. A decrease in the permeability of the soil. This makes certain soils suitable for dam embankments and linings for reservoirs.
4. An increased resistance to erosion.

Compacted soils can be used

1. To make stable ground on which to place the foundations of structures.
2. To provide better support for footings. The compacted soil has a higher bearing value and settlements are reduced.
3. To provide more uniform support for the floor slabs of structures.
4. To provide firm soils for embankments and approach ramps for highways.
5. For the construction of earth dams.
6. For backfills in streets, after pipes or other utilities are placed, and for backfills around footings and around basement walls. It is undesirable for such backfills to settle. Proper compaction prevents settlement.
7. To increase the passive resistance of the soil to lateral loads (including soils compacted around concrete block anchors or around foundations subjected to lateral loads).

22.2 *Mechanics of Compaction*

Compaction increases the weight (dry density) of the soil. This is accomplished by pushing the solid soil particles closer together, and thereby reducing the void spaces in the soil. The void spaces are occupied by air or water, which have no strength.

If the soil particles could be compacted so tightly that all void spaces were eliminated, the resulting solid mass of soil would weigh 160 to 170 lb/ft^3. This is equivalent to the weight of rock. By contrast, soil in its natural condition generally weighs 80–100 lb/ft^3. Since the total weight contains some water, it is somewhat greater than 80–100 lb.

In Fig. 22.1, a cubic foot of soil is represented. However, the soil particles are all pressed together, resulting in 8 in. of solid soil particles. The remainder of the cubic foot is represented as air. This is a bone-dry soil. The soil weight equals $\frac{2}{3} \times 160 = 106$ lb/ft^3.

If the soil had been saturated with water, the air would have been displaced and the top 4 in. of Fig. 22.1 would be water. In this case, the weight of the original soil would be as follows:

$$\text{Soil weight equals } \tfrac{2}{3} \times 160 \text{ lb} = 106 \text{ lb}$$

$$\text{Water weight equals } \tfrac{1}{3} \times 62 \text{ lb} = 21 \text{ lb}$$

$$\text{Total soil weight equals } 127 \text{ lb/ft}^3$$

This generally is referred to as the wet density.

If the dry density equals 106 lb/ft^3 and the water weighs 21 lb, then the moisture content equals

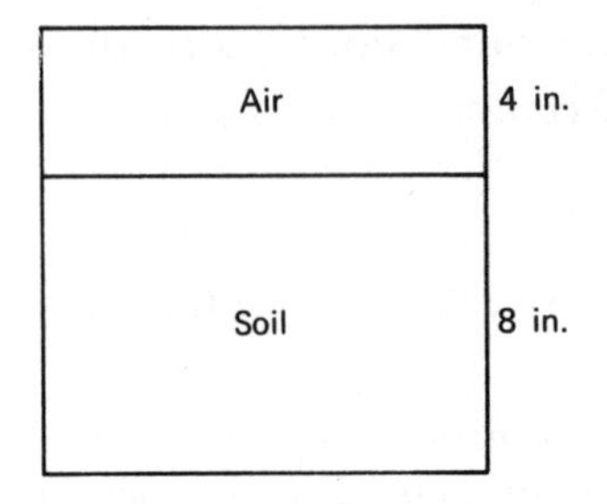

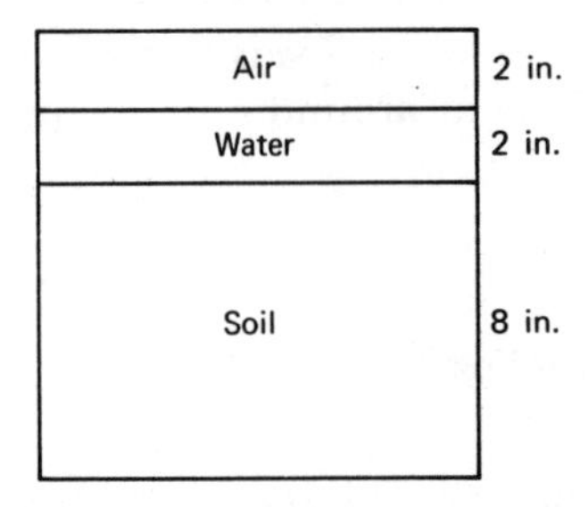

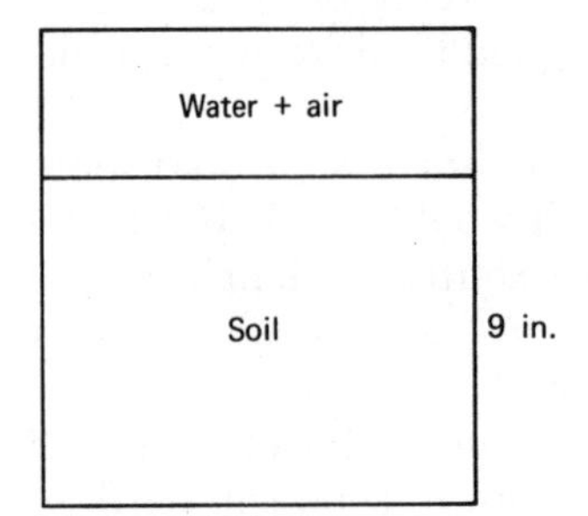

Fig. 22.1 Soil, water, and air relationship in soil mass.

$$\frac{21}{106} = 20\%$$

More commonly, the voids are partially filled with water, as represented by the second diagram in Fig. 22.1. The bottom 8 in. consist of solid soil particles. The next 2 in. are water, and the top 2 in are air. In this case, the soil weight is as follows:

$$\text{Soil weight equals } \tfrac{2}{3} \times 160 \text{ lb} = 106 \text{ lb}$$

$$\text{Water weight equals } \tfrac{1}{6} \times 62 \text{ lb} = 10 \text{ lb}$$

$$\text{Total weight equals } 116 \text{ lb/ft}^3$$

Then moisture content equals

$$\frac{10}{106} = 10\%$$

If the soil must be compacted so that it has a dry density of 120 lb/ft^3, then the volume of solid soil particles would equal

$$\frac{120}{160} = 3/4$$

As shown in the third diagram in Fig. 22.1, the bottom 9 in. could be represented as solid soil particles and the top 3 in. could be water and air. If the soil is saturated, the water content would weigh $1/4 \times 62 = 15$ lb. The moisture content equals

$$\frac{15}{120} = 12\%$$

The amount of water in the soil has a strong influence on the compaction of soil. If the soil is dry, the soil particles are too rough to slide easily. Therefore, it is difficult to rearrange the particles and to compact them more tightly.

If moisture is added, it acts as a lubricant. The particles can slide more easily, and they can be pushed into a tighter configuration.

However, if there is too much water in the soil, the void spaces are too full. There is no way to push the particles together except by squeezing out water. Squeezing out water during compaction generally is difficult, except in very porous soils such as clean sands. However, water can be squeezed out over a long period of time, for example by a structural load.

If the soils become very wet, frequently they are almost plastic. They flow rather than compress under the compaction equipment. Therefore, it is very important to control the amount of water in the soil.

For any soil, and with a specific compaction effort, samples can be compacted at various moisture contents. If several samples are compacted at various moisture contents, the dry density will be different for each compacted sample. Plotting moisture content of the soil against the dry density results in a curve, shown in Fig. 22.2. This curve indicates the percent of water which results in the highest dry density, and is frequently called a moisture-density curve or a compaction curve.

If test strips are compacted on the job site using fill soils containing various moisture contents and identical compaction effort, field density tests would result in a similar curve.

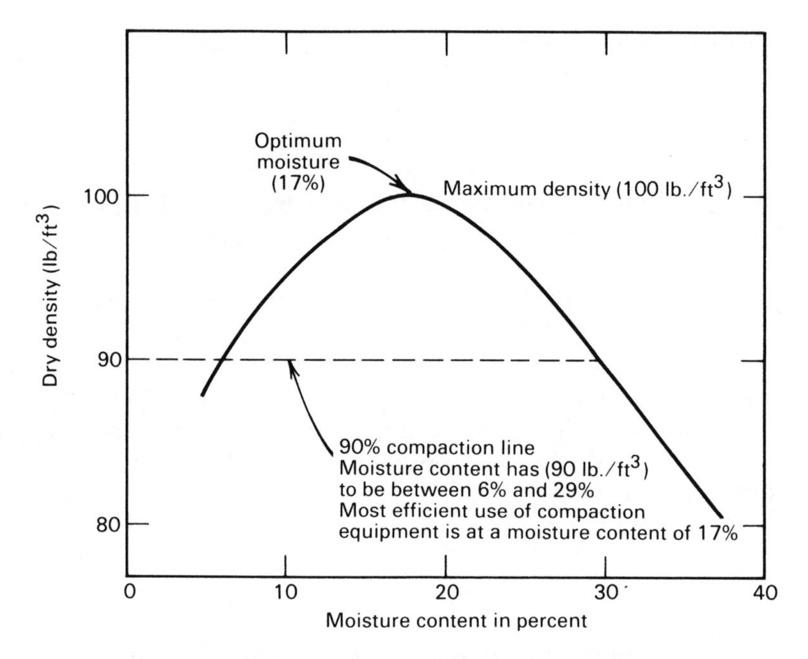

Fig. 22.2 Typical moisture-density relationship or compaction curve.

The purpose of laboratory compaction tests is to attempt to duplicate the compaction on the job site.

The compaction curve indicates the following:

1. The moisture content at which the amount of energy employed produces the most compact soil (optimum moisture content).
2. For a specification requiring that the soil be compacted to 90% of the maximum laboratory density (see Chapter 12, Section 12.9 and also Fig. 22.2), a line can be drawn across the laboratory moisture-density curve. The moisture range where the curve is above the 90% line is the range within which the 90% density can be attained by the compaction energy employed. If the field compaction effort which results from the compaction equipment as commonly employed is well represented by the compaction effort used in the compaction test—as is intended—the range of moisture thus defined is that within which the required density can be attained. Specifications would define a relatively narrow range of allowable moisture content for compaction of the soil. For the example in Fig. 22.2, this might be from about 13 to 20% moisture within the range defined (6 to 29%) for compaction to at least 90% of maximum density.

Apparently, the full pattern for a moisture-density curve, if one could be developed, would be as shown schematically in Fig. 22.3. Completely dry soil existing as individual particles—not cemented—would have a density for a given compaction effort approaching maximum density. As moisture is increased, capillary fringes form in the soil and increase (the "bulking" condition), preventing attainment of maximum density. With further increase in moisture, capillary fringes are destroyed and particles are lubricated, permitting attainment of maximum density at optimum moisture. At still higher moisture contents some pores are completely filled, preventing void reduction, and again maximum density cannot be attained.

This is the apparent complete pattern of moisture-density, but it is somewhat conceptual since the full pattern cannot be developed for a single soil. Free-draining soils, commonly sands with no fines, show the concave-upward shape for the lower moisture range. These are often described as reverse-shaped compaction (moisture-density) curves. The more common concave-downward curve cannot be developed because moisture drains from specimens, preventing compaction at the higher water contents. For most soils only the concave-downward portion of the curve is developed for common moisture ranges, as shown in Fig. 22.2. At the drier conditions that would be necessary for development of the concave-upward portion of the curve, the soil particles range from weakly to strongly cemented and do not behave as particulate materials. As shown in Fig. 22.3, free-draining soils show moisture-density curves like the left portion of the full curve, while all other common soils show the typical curve like the right portion of the full curve or the curve of Fig. 22.2

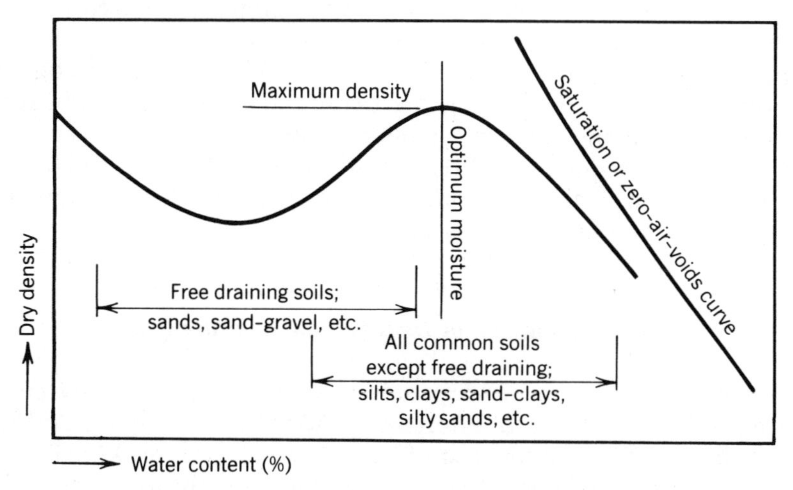

Fig. 22.3 Apparent full pattern for moisture-density curves.

22.3 Energy of Compaction

Increasing the compaction energy applied to a soil results in a higher density or, for a range of moisture contents as shown in Fig. 22.2, in a higher moisture-density curve. A higher curve results, of course, in a higher maximum density, but it also results in a somewhat lower optimum moisture content. In standardized laboratory tests compaction energy can be increased in terms of foot-lbs per cubic inch of soil being compacted. See Chapter 12, Section 12.9, AASHTO Standards T-99 and T-180, or ASTM Standards D-698, *27*, and D-1557, *28*. In the field, compaction energy can be increased by increasing the pattern of stresses induced by the compaction equipment or by increasing the stress repetitions or passes of the compaction roller. Increasing roller weight (drum weight or tire load) causes stresses induced at the surface to reach deeper or attenuate less with increase in depth below the roller. Increasing surface pressure (tire pressure for rubber-tired rollers but only within practical functional ranges) induces higher stresses (and greater compaction energy) at the surface, but without an increase in weight does not significantly increase stresses at depths below the surface. A similar but more complex pattern of pressure or stress increase attends the "walking-out" process for sheepsfoot-type rollers.

The ability of soil to support roller weight or surface pressure is directly related to soil strength. When the soil is overloaded it merely shears and moves aside without increase of density. Where compaction is being effectively accomplished, soil strength is generally increased, permitting some increase in roller tire pressure, and is responsible for the walking-out process. However, the control of soil consistency (moisture content) in relation to roller characteristics is a necessity, or conversely, control of roller characteristics can have very significant impact on compaction effectiveness.

Repetitions or roller passes are directly effective in increasing the compaction energy. But this increase is well established as obtaining only logarithmically. Thus duplicating the densification of a few early roller passes requires ten times as many additional passes. So if two or three passes result in less-than-required density, an added six or eight passes will provide some increase, but if density is much less than required, the addition of more passes is not likely to satisfy needs.

22.4 The Pattern of Densification and Strength

It is common practice to develop only a single moisture-density curve, choosing either the "standard" effort test of ASTM D-698, *27*, or AASHTO T-99 or the "modified" effort test of ASTM D-1557, *28*, or AASHTO T-180 to determine maximum density and optimum moisture, as shown in Fig. 22.2. While this is usually satisfactory, a much better picture of moisture-density and effort is obtained by developing moisture-density curves for three compaction efforts. Usual practice is to compact at standard effort, modified effort, and intermediate effort, as shown in Fig. 22.4. When the line-of-optimums and the saturation or zero-air-voids curves are added to the three effort curves, as in Fig. 22.4, a complete pattern of moisture-density behavior with compaction is shown. The more complete pattern provides a better basis for selection of the most effective compaction moisture for minimum densities specified at some percentage of maximum.

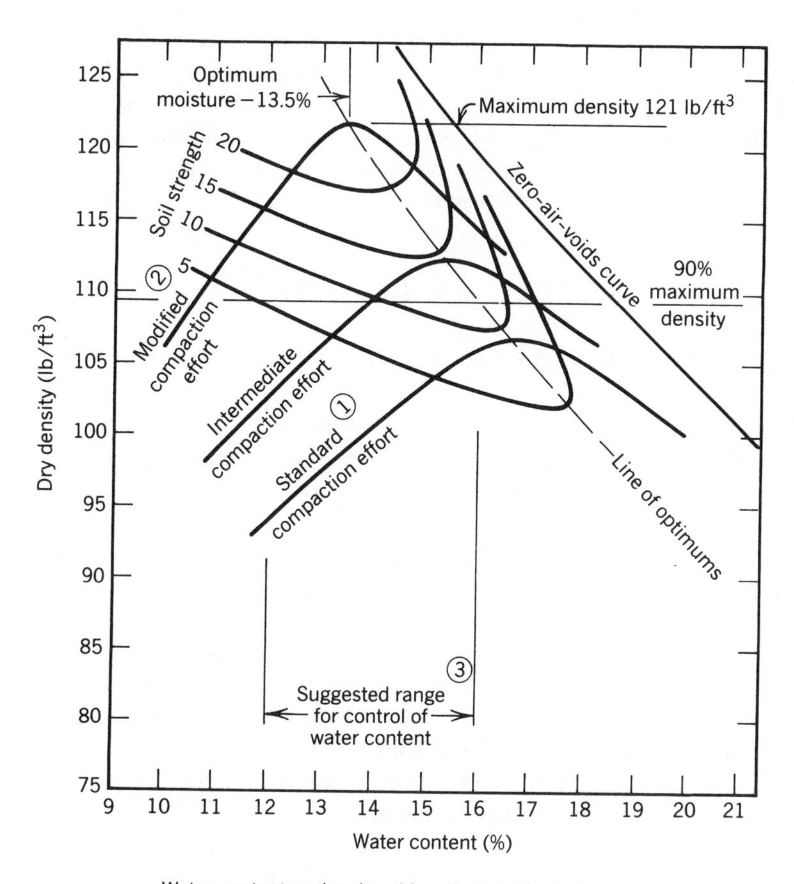

Water content vs density with soil strength contours

① Standard proctor or standard AASHO density.
② Modified proctor or modified AASHO density.
③ For a specified minimum density of 90% modified AASHO maximum density this suggested range for control of water content will reasonably assure the compacted soil will have a strength of 10 or better.

Fig. 22.4 Water content vs. density with soil strength contours.

Of greater importance is the strength pattern to be gained by testing each compacted specimen for strength. The strength pattern can be based on penetration tests, such as CBR or small penetrometer, or unconfined compression tests, vane shear tests, or any that can be employed using the compacted specimens. The resulting strength determinations can be used to develop contours of strength overlaid on the basic moisture-density curves on plots of density versus water content, as shown in Fig. 22.4. The figure gives no indication of how strength was measured, since only the pattern was being represented. The contours of Fig. 22.4 are typical for the CBR-type strength measurements commonly used in design of pavements. They represent subgrade strengths. The base plots are of densities versus the moisture contents at which specimens were molded (unsoaked). The CBR values portrayed are for soaked specimens. Soaked specimens are used in an attempt to represent the lowest subgrade strength to be expected under a pavement which has attained its wettest in-situ condition. A quite similar pattern, but with higher valued contours, would result from tests on the specimens as molded (without soaking).

22.5 Selection of Fill Soil

Job specifications may state that soil will be obtained from the following:

1. Excavations on other portions of the site.
2. Specified borrow pits.
3. Sources to be developed by the contractor.

On-site excavation soil is usually the least expensive. However, if the soil is difficult to work with, the completed and accepted fills may cost more than if better imported materials had been used.

In winter months of frequent rains, soils that ordinarily could be compacted in the summertime may become too wet in the winter rains and become impossible to compact. If the soils are scarified to dry, frequently another rain further saturates the soils.

Therefore, selection of the most suitable soil for construction of a compacted fill depends on the following:

1. Weather conditions during placing of the fill.
2. Suitability of the fill soils.
3. Quality required of the completed fill.

Although a borrow area may be designated as "available to the contractor," it is the contractor's responsibility to verify the suitability of the borrow pit. This includes the suitability of the soil, the extent and volume of suitable soil, the uniformity of the soil, and the presence of unsuitable layers within the borrow area.

Construction of a fill is frequently similar to construction of the poured concrete foundations. The fills are required to be constructed to a high quality. These fills will support foundations or floor slabs. Therefore, fills may be referred to as structural compacted fills. For fills of this quality, construction must be done well to pass the rigid requirements of the specifications and careful testing by an inspection laboratory.

Generally, sandy soils, mixtures of clay and sand or silt and sand, decomposed granite, and some clays are good soils to work with. Other soils can be very difficult to compact, such as silt, siltstone, clean round sand, clay, and plastic clay (such as gumbo or adobe).

22.6 Specifications

Generally, soils used for fill are required to be recompacted to a higher density than their natural density. Most natural deposits of soil have densities ranging from 60 to 90% of maximum obtained by the standard AASHO compaction method.

If sand fill is compacted to the same density as its original condition, the sand performs in about the same manner as the original sand. However, clay has some original structure. Excavation and working the clay change the structure. Some clays are sensitive to reworking and lose some of their strength. Therefore, clays recompacted to their original density may have less strength than the original clay deposit. Usually, most of this strength is regained with time.

However, it is general practice to recompact soils to a higher density than their original natural density. The resulting soils are generally firmer, and consolidate less under load than the adjacent natural soils.

Specifications usually repeat the percentage compaction spelled out in the local building code. In some cases, however, a different percentage compaction may be required to fit a particular purpose. Fills to support heavy footing loads, to be the liner for a water reservoir, or for pavements usually are compacted more than the code requires.

In many cases, sites underlaid by appreciable depths of loose, compressible soils have been considered unsatisfactory for support of heavy structures. However, these soils have been excavated and replaced into the same excavation as compacted fills. In such recompaction, a 20-ft-deep excavation frequently yields enough soil to backfill the excavation with 15 ft of compacted soil. Additional soil must be imported to complete backfilling the excavation. Many heavy structures are supported on such compacted fills.

Specifications should indicate the following:

1. The soils suitable for use as fills.
2. The percentage compaction required for the compacted fills. Structural fills might be specified as 95%; light-duty pavement and other fills outside the building area may be specified as 90%; and yard area fills may be specified as 85%. Occasionally, for airport runways or special conditions, 100% is required.
3. The laboratory compaction test curve, which will be used to establish the maximum density considered to be 100% compaction.
4. The agency responsible for making the tests.
5. The specifications will sometimes spell out the equipment to be on the site for compacting the fill soils; however, the selection of equipment usually is up to the contractor, and the requirement is performance. Performance is measured by testing the fill as it is placed.

22.7 Problem Soils

Occasionally a borrow source is specified which contains soil unsuitable for the proposed fill. The contractor might recognize that the material will not perform as required. Proceeding to use the material hurts the job.

Fill soil is not just dirt—it is a construction material. It deserves as much care in its selection as the concrete, wood, and other materials. The contractor bears a substantial part of the responsibility. Several kinds of soils can be considered as problem soils.

In the northern states, from the New England States through the Middle West and out to Seattle and Portland, good weather for earthwork occurs over a few months of the year. In these months, silty and clayey soils may be used for fills. However, during the major part of the year, silty and clayey soils become wet, are difficult to dry, and can become practically impossible to compact. Once a soil layer of silt or clay is "rained on," it may as well be scraped up and taken off the site. It will not readily dry out. Attempting to scarify and dry this material usually stretches the job out beyond reason. During these wet months, sand is the only steadfast friend of the contractor, even if it is expensive. At least the job can be completed.

Hard silt or shale soils sometimes are excavated and reused as fill. On several such projects, the contractor compacted the soil until it was very hard. Yet, the compaction did *not* come up to the requirement of 90%. Examination revealed that the compacted fill was composed of lumps or "pebbles" of unbroken shale plus a matrix of crushed shale. In the laboratory tests, all the shale was broken down. Recompacted, the density was about 85 lb/ft^3 at 90% compaction. However, the original natural shale formation had a density of about 75 lb/ft^3, which is about 80% compaction. Therefore, in the fill composed of half lumps and half crushed material, the average compaction was the average of 80 and 90%, or about 85%. Such test results called for rejection. When this condition is realized, the requirements should be reviewed with the engineer or owner. Usually, the specifications can be modified, or a more representative compaction test run.

Often soils of inadequate quality can be suitably upgraded by stabilization. Stabilization of soil involves addition of cement, lime, or in some cases asphalt or tar. Other stabilizing agents have been employed but cement and lime remain the most common. The more sandy and less plastic soils can be best stabilized with cement and in some cases can be improved to base course quality. Lime stabilization is more applicable to the more plastic soils, which it improves in texture and to a degree in strength. Sometimes lime is added to improve workability and then cement is added for strength. Dramatic economic advantages have been enjoyed by using lime-cement-flyash in thick stabilized lifts to provide structural layers replacing still thicker conventional structures. Examples are heavy-duty runways at Newark, N.J. and more recently at Houston International Airport, Texas. Stabilization requires special handling and good practice and should not be undertaken too lightly (see Ref. *58*, Chapter 10).

22.8 Equipment for Construction of Fills

22.8.1 Rollers

A great variety of rollers are manufactured for compaction of soil. These include:

Sheepsfoot rollers.
Wobble wheel rollers.
Straight wheel rollers.
Super compactors.
Flat wheel road rollers.
Grid rollers.

Sheepsfoot rollers come in a variety of sizes. The size usually is indicated by the roller's width and diameter. A 5 by 5 roller is 5 ft in diameter and 5 ft wide. A 4 by 5 roller is 4 ft in diameter and 5 ft wide. Common sizes of rollers are 4 by 4 ft, 5 by 5 ft, and 6 by 6 ft.

The drum usually is loaded with water, or with sand and water. The roller is unloaded for transportation between jobs and loaded for use on the job site. A roller may be used empty or only partly loaded if it is starting out on a moderately soft subgrade.

Feet generally are about 7–9 in. in length, and the end of a foot has a bearing area of 5–9 in.^{2}. There are several foot shapes.

The pressure on the bottom of each foot is figured by dividing the total load of the roller by the area of all of the feet in one row, assuming only one row of feet is on the ground at one time. Typical foot pressures are in the range of 200–400 psi.

Sheepsfoot rollers work well on sand containing silts and clays. They work moderately well in silts, and work satisfactorily in clay as long as the moisture content is at or below optimum. Sheepsfoot rollers may not be too effective in relatively clean or uniformly graded sand.

22.8.2 Vibrators

Vibrating compactors have become very popular. They include

Vibrating drum rollers.
Vibrating flat place compactors.
Vibrating sheepsfoot rollers.

Vibration is imparted to rollers or plates by pairs of counterrotating eccentric masses or by hydraulically actuated reciprocating masses. These produce dynamic forces which successively reduce and then add to the static weight. Commonly, this reduces static weight to near zero and then increases it to nearly double the original static weight with a consequent increase in the compacting effort.

The vibration imparted to the soil can reduce the forces between individual soil grains, permitting them to move into denser formations. Thus vibrating compactors are most effective in cohesionless soils such as sand, sand and gravel, or mixtures of sand or gravel containing some silt. The more plastic, clayey soils are sustained by cohesive bonds between soil particles which are not readily disrupted by the stress variations induced by vibration. Thus vibratory compactors are less effective to largely ineffective in clays, silty-clays, sandy clays, or clayey gravels.

Some lower frequency vibrators are more effective in plastic soils when their reciprocal force variation is effective as individual tamping type compaction, but higher frequency compactors most effective in nonplastic soils are more commonly available.

Vibrating plate compactors are used primarily for compacting in confined areas.

Photographs of some compactors are shown in Fig. 22.5(*a*) and (*b*).

22.8.3 Hauling Equipment

Hauling equipment sometimes is used to compact fills. Frequently, large hauling units such as scrapers or dumptrucks are routed over fills to obtain additional compaction. In this case, the hauling equipment does not follow the same path across the fill each time. Instead, it follows new paths on each trip. These heavy units are efficient in compacting many soils and may be pressed into service as compaction equipment from time to time. On smaller construction jobs, the use of loaded dumptrucks as compaction equipment is fairly common.

Fig. 22.5 (*a*) Typical compaction equipment.

Bulldozers are sometimes used to compact the soil. Provided the soil is properly moistened and spread in thin layers, usually 3 to 4 in. thick, a bulldozer can in many cases achieve suitable compaction of the soil. The bearing pressure of the bulldozer tracks generally is low, 6 to 10 psi. Therefore, they are not suitable for achieving a high degree of compaction. By contrast, the pressure imposed by the feet of a sheepsfoot roller generally is around 200 to 400 psi.

22.8.4 Tampers

For smaller fills and backfills, space limitations prevent the use of rollers. Tampers are used. Commonly used tampers include

Barco compactors.
Ingersoll-Rand "simplex" air tampers.
Wackers.
Del-Mag thumpers.

For efficient compaction with hand-operated tampers, fill layers usually are about 4 in. in thickness. This might be reduced to 3 in. for clay and

Fig. 22.5 (*b*) Typical compaction equipment.

increased to 5 or 6 in. for sandy soil. Tampers usually are not very efficient in compacting clean sands.

22.9 Compaction of Existing Soil In-Place

The previous sections described construction of compacted fills as they are placed. Sometimes, it is necessary to compact an existing body of soil. This could be a loose fill or loose natural soil.

22.9.1 Vibroflotation

The vibrofloat operates in a manner somewhat similar to a concrete vibrator or "stinger" (see Fig. 22.6).

Vibroflotation is used to compact loose sands. It can compact to depths of 30–40 ft. The vibrofloat is inserted into the ground, vibrates back and forth in its hole, and compacts the surrounding soil. As the surrounding soil is pushed sideways, sand is dropped down the hole to make up the lost volume. Water is injected during this process.

The vibrofloat works best in loose, clean sand, or soil that drains rapidly. When the content of silt or clay exceeds about 10%, the process loses its effectiveness. The process works below or above water level.

Under a large mat foundation, the vibrofloat may be inserted in a pattern or spacing of 6 ft apart, or perhaps 8 ft or even 10 ft apart. Under an individual footing 6 by 6 ft in size, the vibrofloat may be inserted three times on a spacing of 6 ft center-to-center. Generally, sands can be compacted to 90% of modified AASHO.

In bidding on vibroflotation compaction work, it is customary to guarantee that the sand will be compacted to a certain percentage of "relative density." Relative density is different from percentage compaction as normally

Fig. 22.6 (*a*) Vibroflotation machine.

Fig. 22.6 (*b*) Vibroflotation machine.

used in fill work. To illustrate the method of calculating relative density, assume that a particular sand can be compacted to a maximum density of 120 lb/ft^3 and can be placed at a loosest or minimum density of 70 lb/ft^3. Sand in its loosest state has a relative density of 0 and in its densest state has a relative density of 100%. If the maximum density of the sand is 120 lb/ft^3, the actual compacted density in the field is 100 lb/ft^3, and the loosest condition for the sand is 70 lb/ft^3, the relative density would be calculated as

$$100\% \times \frac{120}{100} \times \frac{100 - 70}{120 - 70} = 72\%^*$$

It is common to specify relative density values ranging from 70% up to 80 or 85%.

Sand compacted to a relative density of 70% will support bearing pressures of 6000 lb/ft^2, while sand at 80% relative density is good for about 8000 lb/ft^2.

* The relative density is defined in terms of the maximum, minimum, and actual void ratios. The void ratio is defined as the volume of voids divided by the volume of solids. Extension of these definitions leads to the equation presented. For more detail see any of the following references: *3*, *13*, *25*, or *49*.

Clean coarse sand responds very well to vibroflotation; an extremely dense and compact fill can be achieved. Marginal results are obtained in fine sand with some silt.

Soil gradation determines the effectiveness and feasibility of this process. Other factors are involved, of course, but gradation appears to be of primary importance. The Vibroflotation Foundation Company usually works as a specialty subcontractor.

22.9.2 Piles

Sometimes piles are driven into loose sandy soil to compact it. Ordinary piles could be used, such as timber piles. However, the method usually used is called "sand piles," because each pile consists of a column of sand. The process usually is as follows:

> Drive a pipe into the ground. There is a plate "loose fitted" on the base of the pipe, so that the pipe remains empty. Fill the pipe with sand. Pull out the pipe, leaving the sand and bottom plate behind. These piles may be placed at intervals like 8 ft on centers each way, and sometimes are 40 or 50 ft long.

Such piles also may be used for another purpose, to permit rapid drainage and settlement of soft mud.

Surcharge loading. Sometimes fill is temporarily stockpiled on a site to consolidate the underlying soil. The increase in relative compaction generally is small (see Chapter 19, Section 19.7.1).

Pressure injection. See Chapter 28.

22.9.3 Large Vibrators

Large vibrating rollers now are available which can compact granular soils to depths of several feet. Frequently, such compaction equipment is suitable for densifying a deposit of existing soil to a sufficient depth to provide adequate support for spread footings [Fig. 22.5(*a*) and (*b*)].

In a recent test, it was found that a large vibrating roller, generating a downward impact force of approximately 40 tons, was able to compact a silty fine sand soil in a zone from 3 to 7 ft below grade. The increase was as much as 20% in relative density. At a depth of 7 ft, the relative density was increased as much as 5%. The top 3 ft were not compacted, and a smaller roller was required to compact this zone.

22.9.4 Explosives

A number of sites underlaid by loose sandy soil and having a high water level have been compacted by setting off high explosives within the soil (see Ref. *59*).

22.9.5 *Dewatering*

Lowering the groundwater level temporarily at a site is used frequently as a means of consolidating loose deposits of hydraulic dredged fill, or other silty or sandy soil. Dewatering can be done by digging perimeter drainage ditches and installing sump pumps, or by installing well points or wells.

Dewatering has two effects on the soil:

1. The water is removed more quickly than by normal drainage.
2. The soil is no longer "buoyed up" by water, and effectively becomes heavier. This greater effective stress in the soil causes more settlement to occur. If the water level is temporarily pulled 10 ft below permanent water level, the temporary increase in effective stress in the soil is about 400 lb/ft^2. This is equivalent to the weight of most two- or three-story buildings.

22.9.6 *Selection of Compaction Equipment*

The type of soil proposed for use as fill material will be the primary factor in deciding on compaction methods and compaction equipment. Some specifications require that soils that are excavated be hauled off the site and disposed of, and that selected fill materials be hauled in. However, in most contracts, it is anticipated that soils excavated on portions of the site will be used for fill or backfill in other portions of the site. Usually, these soils will vary with depth and also with location.

From the standpoint of economics of operating the job, the contractor should be sure he or she knows the construction materials to be used.

The first step in getting compaction is to get the right moisture content into the soil. Wet silt and clay usually are difficult to dry out. Spreading and breaking up the soil is slow and expensive, requiring equipment designed for the purpose, such as Pulvi-Mixer-type machines. Dry silt and clay soils also are difficult to work moisture into. Presprinkling in the borrow areas can save time. Some silty soils are powdery, and may require mixing along with presprinkling in the borrow area. Rainbirds and similar sprinklers, with temporary quick-connect pipe systems, usually are used.

Sandy soils generally are not a problem for moisture-conditioning and for compaction. They compact well with vibrating rollers and wheel rollers.

Silty and clayey soils generally can be compacted efficiently with sheepsfoot rollers. The feet avoid the formation of a surface crust. The feet reach down and compact the lower portion of each lift of soil. In some soils, the sheepsfoot roller will "walk out," meaning that the lower and intermediate portions of the fill lift are compacted and perhaps only the top 2 or 3 in. of soil are still loose. However, in all cases, there is some loose soil on top of each lift.

The ideal method of selecting equipment for a project is to set up a test strip on the site and to experiment using various pieces of compaction equipment with various thicknesses of soil layers. Also, try other variations, such as two or three different moisture contents of the soil; for example, if optimum is 14%, try 12%, 14%, and 16%.

22.9.7 *Compaction Procedure*

Probably the most important factor in compaction of fills is the moisture content of the soil. If the moisture content is at or near the optimum, compaction generally is relatively easy. If the range of moisture content is a few percentages above or below the desired moisture content, compaction still is possible, but more effort is required. When the moisture content is several percentages above or below optimum and is outside the "limiting percentages," compaction is extremely difficult or perhaps impossible. The concept of limiting percentages of moisture content is shown in Figs. 22.2 and 22.4.

It may seem a waste of time and money to struggle to get the moisture content close to optimum. However, the greater the variance between field moisture and optimum, the greater the compactive effort needed. More passes with equipment also cost time and money, and increase the chances of fill lifts being rejected for inadequate compaction (see Fig. 22.7).

The lifts of fill should be uniform in thickness. The most efficient thickness for the type of compaction equipment being used can be determined by construction of a test fill strip. The most efficient fill thickness may vary from 2-in. lifts where tampers are used, to 6- to 8-in. lifts for sheepsfoot rollers, to 10- to 12-in. lifts for vibrating rollers on sandy soils, to 12- to 14-in. lifts for heavy compactors. Loose lifts 8–10 in. thick are fairly common, and compact down 2 or 3 in. to a compacted thickness of 6–8 in.

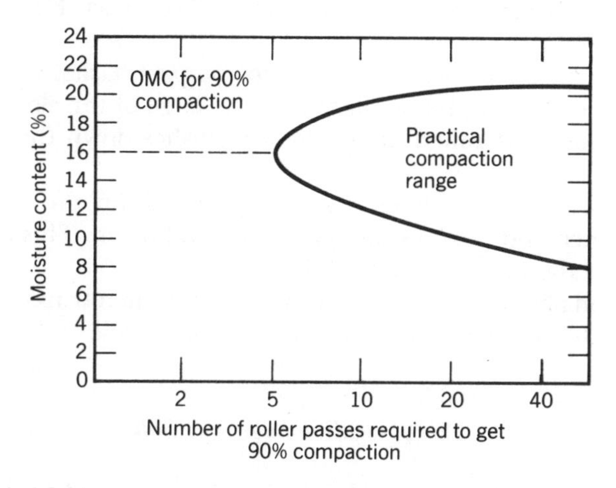

Fig. 22.7 Typical field compaction pattern. Note that at 22% moisture content compaction to 90% is not possible no matter how many roller passes are made.

During compaction of fills, it is common practice for the owner to arrange for tests to be made of the completed layers of fill. Where these tests indicate that a layer is not satisfactory (has failed), the contractor should check on the details of the test results. If the moisture content of the soil is too high or low, the poor results may be attributed to improper moisture content and indicate the corrections to be made. If the soil is at the optimum moisture content, it is possible that not enough passes have been made or that the wrong type of equipment is being used. If the test results are irregular, for no apparent reason, the soil being imported may not be uniform. The soil may change, requiring additional testing of the soil to establish a new compaction curve. It is in the contractor's best interests to understand the theory and test procedures involved in soil compaction. Test results can be used by the contractor to get the required compaction at minimum cost.

When compaction is difficult, and after many passes the fill still does not quite make 90%, the question is what to do. Frequently, contractors will try a bigger and heavier roller, which may help. If a pneumatic-tired roller is used, increasing tire pressure may help.

Since heavier rollers, using higher tire pressure where pertinent, apply greater energy (for equal passes) to the soil lift being compacted, the optimum moisture (OMC) for this greater energy will be lower—commonly by about 3%. Thus if the fill is on the dry side of the OMC for the standard test or the specified effort, the heavier roller will help. If, however, the fill is wet of the OMC for the specified effort, heavier roller compaction will be detrimental (see Fig. 22.4). In this case it will be necessary to dry the soil or resort to chemical stabilizing or drying as described in Section 22.7 (also see Ref. *58*, Chapter 10).

Sometimes contractors try a "sandwich" method of constructing with wet soils. Alternating layers of wet soil and dry soil are placed. This may dry out the wet layers enough so that they will compact. However, success varies.

Fills placed on slopes are difficult to compact. Usually, the fill layers can be compacted to within about 2 ft of the edge of the slope. The outer 2 ft remain loose. If rolled upon, the soil pushes down the slope rather than compacting.

Frequently, slope rolling is done up and down the slope, using a bulldozer winch and cable to run a roller up and down. This compacts the outer 6 in. of the fill. The other 18 in. remain loose.

Occasionally, the slope is "overbuilt" by 2 ft, then trimmed back. This removes the loose soil and exposes the compacted soil.

22.10 Filling over Soft Soil

In many cases of land development, good quality fills are placed over swampy or marshy land. Starting the job can be very important. It is not possible to plunge out onto the soft mud and lay and compact conventional layers of fill soil. The mud may fail, creating mud waves and making the soil even softer.

In almost all cases, it is desirable to maintain the integrity of the underlying soil and to take advantage of its limited strength. This may be done by pushing out a first layer which is relatively thick, perhaps 12–18 in. thick. This first layer should be pushed out with very small and light equipment. After the first layer is placed, it should be rolled with light equipment. The equipment suitable for such rolling can be determined by experimenting.

Succeeding lifts can be placed with larger equipment, and larger compaction equipment can be brought onto the succeeding lifts to obtain increasingly better compaction as thickness of fill builds up. This method of operation is shown in Fig. 22.8.

In many cases, fills are to be placed in marshy areas with heavy growth of marsh grass or other vegetation. It would be difficult to go into the marshy area with excavating equipment to strip out the grass and vegetation to reach clean soil on which to start the fill. One approach is to use light mowing equipment, or even hand labor, to cut off the vegetation close to the ground. Burning also may work in some cases. The remaining stubble provides some strength on which the first layer of fill can be placed. At first there might be some reluctance to leaving some organic material in place. However, a few inches of organic material may be much less compressible than two feet of soft soil churned up and diluted with water.

Another possible method of filling is called mud displacement. Frequently, if the mud is not too deep, the contractor elects to displace the mud. In this case fill is built up to a considerable height. Its weight causes the mud to flow outward. The fill settles down on the firmer soil under the mud. The mud wave must be kept moving; if it "freezes" it has to be dug out or sometimes is blasted to get it to start moving.

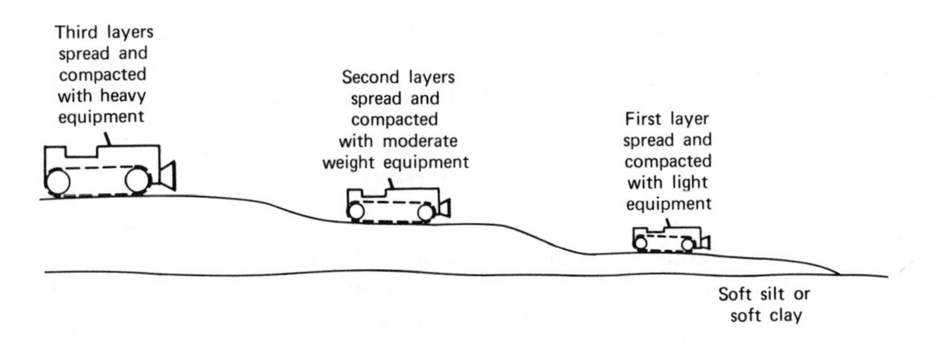

Fig. 22.8 Procedure for construction of compacted fill on bed of soft soil.

More recently geotextiles and geomembranes have been used to prevent penetration of the fill into the soft underlayer and to control displacement of the soft underlayer. This will normally be a matter of design of the filling rather than a contractor option, but in cases where a temporary road or platform is needed, the use of fabrics or membranes can limit both disruption of the underlying layer and thickness of overlying lift needed to provide the desired fill structure.

22.11 Protection from Weather

22.11.1 Rain

Rain is a problem to almost all compacted fill operations. If the soils used are silt or clay soils, or the work is being done in a cloudy, damp season in which evaporation rates are slow, rain can cause delays and may become a disaster.

Some methods used to protect the surface of the fill include the following:

Quickly blade-smooth the fill surface, and smooth-roll the surface to make it shed water and reduce the amount which soaks in.

Provide drainage to get the rainwater off the compacted fill area as quickly as possible. Also, the borrow area may need to be "closed up" to avoid saturation of the soil.

Provide perimeter ditches and pumps to collect rainwater and then pump it off the job site.

Occasionally, on small fills or repair work, special treatments may be used, such as the following:

Sheets of plastic such as polyvinyl-chloride can be bought in wide sheets from builder's supply stores. These sheets can be held down by placing sandbags on them at regular intervals.

Sometimes lime is added to the soil and scarified and bladed in to absorb moisture and make the soil more workable. The amount needed varies with soil conditions, but an average is in the range of 4–6%.

22.11.2 Freezing

Clay and silt soils are difficult to compact in the winter when freezing occurs. These soils hold moisture, which freezes and becomes hard. This soil cannot compact when it is frozen hard. The next summer, however, when the soil thaws, the frozen layers will still be uncompacted and loose. This results in settlements and soft fills. When freezing occurs, the top layer of frozen soils should be stripped off the site each morning before new fill is placed.

As an alternative, it may be worthwhile to cover the fill area each night with a blanket of soil which would act as a protective surface. This soil would be bladed off the next morning and stockpiled for reuse again that night.

Soils can be compacted in air temperatures below 32° provided the soil itself is not frozen. This requires good planning.

When fills are completed to desired rough grade, they should be covered by a protective layer of soil, or other material such as straw or hay, until other permanent construction can be started. Considerable research in winter construction has been performed by the U.S. Army Corps of Engineers and others (see Ref. *60*).

22.11.3 Drying

In dry desert areas, the soils may dry out rapidly, creating a problem in maintaining the desired moisture content. Also, fills left overnight may dry up considerably on top and possibly cause surface shrinkage cracking. It may be necessary to install rainbirds or other sprinkling systems which can be operated by clock to keep the fill surface wet over weekends or to keep completed fills properly moistened prior to pouring of concrete floor slabs.

22.12 Boulders

On projects in which balanced cut-and-fill quantities have been developed, boulders may be a problem. Hauling the boulders off-site is expensive, and it may be difficult to find a proper disposal area. Also, the loss must be made up by importing fill. Therefore, consideration must be given to using the boulders in the proposed fills.

The problems resulting from the use of boulders are as follows:

- If the boulders are too close to the final grade, they may interfere with the proper performance of foundations if the foundations are supported partly on soil and partly on boulders.
- Boulders may interfere with later trenching for pipelines.
- Boulders may interfere with proper compaction of soil, resulting in loose soils in boulder areas and possible future settlement.

It would be desirable to designate areas, such as parking areas, in which no structures would be built or pipelines constructed. Also, a review of grades might indicate areas of deeper fills, where it would be safe to dispose of boulders in the deep portion of the fills.

To compact fill soils around boulders, one method is to place the boulders in rows. Fills can be placed along each side of the row of boulders, with the lifts coming up evenly on both sides.

Some hand tamping work may be necessary in tight spots.

22.13 Inspection and Testing

22.13.1 The Inspector

Almost all compacted fills are inspected and tested during placement by a representative of the owner. The representative, or inspector, may be an employee of the owner or may be a soils engineer or a soils laboratory hired by the owner.

The contractor can obtain information from the soils inspector. When tests indicate the fill is not properly compacted, one should find out why. Is the moisture too far off? What percentage compaction actually was obtained?

Though considered by many to be an unsound practice, it is more and more commonly being required that the contractor provide for quality control of compaction he or she is hired to perform. The common argument against this practice is that the contract compliance inspection must be by a representative of the owner since contractor control can too easily lead to ineffective and uncaring operations and even to distortions and cover-up. The argument for contracting the quality control inspection is that it is a necessary function requiring significant cost. It is nearly always too large a function to be expected from an owner's soil engineer representative, who is supported by costs separate from the basic contract.

In either case it is to the contractor's advantage to understand and control the compaction process.

On most compacted fill projects, the soils engineer will place a field laboratory on the site and have an inspector full-time on the job. On small jobs, the soils inspector may come to the site intermittently, as fill is being placed, and may operate with scales and equipment in the back of a truck. The purpose of inspection is to do the following:

1. Check the soil being used for fill.
2. Observe the compaction operations for complete coverage of each lift of fill.
3. Test the fill layers as they are completed. Notify the contractor whether the tests passed.
4. Review the problem with the contractor if any lifts fail. Indicate whether high moisture content, low moisture content, nonuniformity of compaction, or what other conditions cause the failure. This will permit the contractor to decide how best to correct the situation.

In most cases, a city, county, or federal agency is involved in overall building permit and plan checking, which requires that the soils engineer submit reports to the governmental agency regarding satisfactory completion of the fill.

Fill tests are made by several methods:

1. The sand replacement or "sand cone" test. In this test, a sample of completed fill is dug out. The volume of the resulting hole is measured, using calibrated sand. This test is described in more detail in the following section and in ASTM D-1556, *61*.
2. A balloon testing device. A hole is dug into the fill. The volume of the hole is measured by the balloon apparatus, *62*.
3. Nuclear testing. Various nuclear radiation devices have been developed to measure the density and moisture content of the soil, *63*. The manufacturers of such equipment are listed in Ref. *64*.
4. A test in the field laboratory of a core of the fill soil obtained by pushing or driving soil samplers into the ground, *65*.
5. A probe with a calibrated device, called a penetrometer. One type of penetrometer is the Dutch cone. Resistance to penetration is correlated with previous experience as a basis for estimating the soil compaction.

22.13.2 Sand Replacement Method

The sand replacement method is slow and tedious but is the most commonly used method. The steps in performing this test are as follows:

1. *Calibrating the Sand.* Dry sand of a known density is used to measure the volume of soil removed from the fill. The sand used is "Ottawa" sand or other sand of uniform grain size. When poured at a constant rate of flow from a constant height, the sand will deposit at a uniform density. The density of the sand is determined by carefully pouring it into a compaction cylinder having a known volume. When the cylinder is full, it is struck off level and weighed. The weight of the cylinder is then subtracted to obtain the weight of the sand. The weight of sand is divided by the volume of the cylinder to get the density of the sand.
2. *Performing the Test.* At each location to be tested, the upper 3–6 in. of loose soil is removed. Frequently this is done with the blade of a bulldozer. A level surface is made on the fill. A metal ring is placed on the fill. A hole, usually about 5 to 6 in. in diameter, is dug through the test ring and into the soil. The hole is dug to a depth of about 5 to 6 in., which will provide approximately 5–10 lb of soil. The soil removed is placed in a can with a tight lid to avoid loss of moisture. Every crumb of soil is dug out and placed in the can.

   The test hole is filled with the calibrated sand. The sand is taken from a can or jar containing a specific weight of sand. Therefore, the weight of sand used to fill the hole can be determined by weighing the jar and remaining sand. During pouring of sand into the hole, compaction equipment which might cause vibration of the ground should not be operated close to the test.

3. *Calculating the Test Results.* The soil removed from the test hole and the remaining calibrated sand are taken to the laboratory. The soil removed from the hole is weighed. The calibrated sand is weighed and subtracted from the original weight of sand to determine the amount used to fill the test hole.

   The fill soil wet density is calculated as

$$\frac{\text{Weight of soil}}{\text{Weight of sand}} \times \text{calibrated density of the sand}$$

If the sand-calibrated density is 90, then the formula becomes

$$\frac{\text{Weight of soil}}{\text{Weight of sand}} \times 90$$

This answer is in pounds per cubic foot. Typical soil wet-density values are 100–130 lb/ft^3.

EXAMPLE. Weight of soil = 10 lb
Weight of sand used = 8 lb
Calibrated density of sand = 90 lb/ft^3

Fill soil wet density:

$$\frac{10}{8} \times 90 = 112.5 \text{ lb/ft}^3$$

The dry density of the fill soil is determined by placing all, or a portion of, the fill soil in an oven and measuring the loss of weight when the soil is bone-dry. In field laboratories, a small electric plate or gasoline stove can be used with fairly accurate results. A few checks should be made with an oven drying at 105°C. The percentage moisture in the fill soil is calculated as follows:

Step 1: Weight of moisture lost during drying = wet weight of soil minus dry weight of soil.

Step 2: Percent moisture = (this answer is in percent)

$$\frac{\text{Weight of moisture lost}}{\text{Dry weight of soil}} \times 100$$

EXAMPLE. Weight of soil = 10 lb
Dry weight of soil = 9 lb
Moisture lost = 1 lb

Percentage moisture:

$$\frac{1}{9} \times 100 = 11.1\%$$

The dry density of the fill soil is calculated as follows:

$$\text{Dry density} = \frac{\text{wet density of soil}}{\text{percentage moisture} + 100}$$

EXAMPLE

$$\frac{112.5}{11.1\% + 100\%} = \frac{112.5}{1.111} = 101 \text{ lb/ft}^3$$

Testing of fill is not a very accurate procedure. Also, it is possible to sample at an unrepresentative spot. Statistical methods sometimes are used on larger projects to make logical decisions from test results. On this basis, an occasional failing test can be accepted if the daily averages are consistently above the specified percentage compaction.

23

Floor Slabs

23.1 Slabs on Grade

23.1.1 Subgrade

The floor slabs for many large industrial buildings, particularly where large areas are involved, are supported directly on the ground.In many parts of the country, residential housing, schools, and commercial structures also use floor slabs poured on grade.

For such buildings, the underlying soil becomes a structural part of the building. Detailed information regarding the quality of the subgrade soil is necessary.

The subgrade "structural soils" should be prepared properly. This may include removal of topsoil or unsatisfactory soils, and may include proofrolling of the floor area, as described in Chapter 21, Section 21.1.

For floors that support heavy loads or are subjected to heavy traffic of forklift trucks, it is common to place a base course under the floor slab. The base course may consist of crushed rock, sand and gravel, cement-stabilized sand, decomposed granite, or other suitable material, which, after compaction, experiences relatively small deflections under load.

23.1.2 Settlement

The columns and walls for such buildings are supported on independent foundations. These may be spread foundations carried down to firmer soils. However, in many cases the foundations are placed only 2 or 3 ft below the floor slab.

The columns may be loaded with 30 or 40 kips per column or with as much as 200 or 300 kips per column. During construction, these foundations will settle as the dead loads of the building are applied. In the case of reinforced concrete industrial structures, a large proportion of the total load may be dead load of the structure.

Usually, settlements of the floor slab are relatively small, compared to the foundation settlements. Therefore, the construction schedule should be such that the structure is essentially completed and dead loads are on the building columns before pouring of the floor slab. In addition, frequently it is desirable to leave a separation between the floor slab and the columns. Therefore, future settlements can occur with the columns sliding through the holes in the floor slab.

If the floor is used to support heavy storage loads, particularly pallets stacked 15 or 20 ft high with canned goods, tin plate, or other extremely heavy materials, the floor slab loading can be several hundreds, or even thousands, of pounds per square foot. Usually, such storage is placed in islands, with aisles between. Therefore, the soils underlying the floor slab are subjected to heavy stresses. This may cause settlements, differential settlements between storage areas and aisles, and possibly a failure of the soil. Such floors may settle 6 in. or 1 ft.

In many parts of the country, floors for residences or one-story buildings may also serve as the foundations for the structures. In these cases, the floor usually is thickened, or is thickened under the perimeter and interior wall lines. Settlements usually are small because loads are small. The main requirement is that the subgrade soil be stable. Unstable soil may be

Soft soil or mud.

Expansive soil, gumbo, or adobe.

Collapsing soil.

Loose fill or dumped debris.

Soil loosened by tree removal.

In *all* such cases, the soil must be compacted, "stabilized," or removed before construction. After construction, repairs are expensive and lead to many lawsuits.

In hot, dry weather, soil drying and cracking occurs in many floor excavations which are left open for a long period of time. Drying and cracking of the soil should be prevented. It will not result in settlement, but on the contrary, will cause the floor slab to be uplifted when the soil later regains its moisture. Uplifting may be uneven, causing cracks in the floor. Methods of protection are described in Chapter 15, Section 15.2.1.

23.1.3 Underground Utilities

When the grading contractor prepares the site and completes a fill for the building area which is composed of properly compacted materials, he or she leaves the site with a feeling of satisfaction of a job well done. However, as soon as the grading contractor leaves, a multitude of other subcontractors enter the site. They bring trenching machines and soon have the site torn up in a maze of crisscrossing trenches. Sewer lines, water lines, fire lines, gas lines, electrical conduits, steam lines, and a variety of other utilities are placed in the ground. Then these trenches are backfilled. It is this backfill which must support part of the floor slab.

Usually, the plumbing and other subcontracts are separate from the grading contract, and the requirements for compaction may not be similar. In some cases, utility backfills are considered a nuisance to the plumbing contractor, and the backfill reflects this lack of concern.

When the floor slab settles later and cracks, the original grading contractor may be called on to repair his or her "improper work." While it should be the responsibility of the architect or primary designer of the structure to make sure that utility backfills are as good as the original site grading, it is also in the best interest of the general contractor to make sure that this situation is properly considered.

23.1.4 Heavy Floor Loads

It is fairly common for floor loads to be in the range of 500 to 1000 lb/ft^2 and occasionally to amount to several thousand pounds per square foot. Floor settlements may amount to a few inches up to 6 in. to perhaps 1 ft.

Therefore, the design of the floor slab support becomes as important as the design of the foundations for the building itself. The future settlements of a floor slab can be estimated using procedures described in Chapter 19, Section 19.2.

23.1.5 Moisture Barriers

Soils underlying floor slabs contain some moisture. The amount of moisture depends on the type of soil, depth to groundwater level, site drainage, and rainfall. However, most soils contain at least 10 to 20% moisture. Even in the desert areas, soils which appear to be bone dry contain 4 or 5% of water by weight. This represents some 2 or 3 quarts of water per cubic foot of the dry-appearing soil, or 1 or 2 gallons of water for more normal soil. This water may pass through the floor slab either as moisture or as vapor. On reaching the surface of the floor slab, it evaporates. If a fairly substantial amount of water flows up through the floor slab, it may leave behind white salts, commonly called efflorescence.

However, in office areas or in houses with slabs on grade, it is likely that the floors will be covered with tile surfaces, such as vinyl-asbestos tile or plastic tile. These tiles are fastened to the concrete with adhesives or mastic-type glues. Also, some concrete floor slabs are covered by rugs.

The moisture and water vapor moving upward through a concrete floor slab will collect under the asphalt or plastic tiles. (This can be observed whenever old slabs or pavements are removed. The soil underneath is wet.) Once sufficient water has collected, the tiles will loosen or "pop off." Moisture collecting under wool or fabric carpeting may cause the carpets to mildew.

Therefore, moisture barriers are commonly placed under floor slabs to stop the upward flow of moisture and water vapor.

It has been common practice to place a layer of free-draining sand or gravel immediately below the floor slab. This material has no "capillary rise" and therefore prevents moisture from reaching the bottom of the floor slab.

Studies have indicated that part of the moisture passing up through the floor slab is water vapor. The layer of granular soil or crushed rock apparently will not block the flow of water vapor.

Development of plastic sheets of good quality, strength, and low cost has promoted the use of plastic membranes for moisture barriers. These membranes are placed below the floor slabs. In many cases, the membranes are covered with 2 or 3 in. of sand. The sand covering serves two purposes:

1. It provides protection to the plastic membrane when the wire fabric or reinforcing steel is placed for the slab, and during pouring of the concrete slab.
2. It can absorb from the floor slab excess water developed during curing. This solves the problem of curing of concrete poured on plastic membranes.

By contrast, some feel that the concrete slab *should* be poured directly on the plastic sheets. This avoids the sand "reservoir" below the floor slab. The concrete curing problem is handled by careful control of the concrete mix and curing.

23.1.6 Basement Floors

The groundwater level may be very close to the elevation of the basement floor slab. During rainy seasons, the groundwater may even rise above the level of the floor slab. Therefore, additional protection is needed.

Underdrains may be placed under the floor slab. Usually, they are perforated plastic pipes or porous concrete pipes, placed in a bed of properly graded sand and leading to a sump in the basement floor. As water flows through the pipes and into the sumps, it is pumped out periodically.

The basement floor slab and walls require waterproofing. The waterproofing usually consists of sheets of asphaltic felt mopped with asphalt or tar, underlying the floor slab.

Occasionally, relief holes or relief valves are placed in a slab if it is anticipated that the groundwater level might suddenly rise and develop a hydrostatic uplift pressure on the bottom of the slab.

23.2 Slabs at Railroad Car Height

23.2.1 Subgrade

Many industrial buildings have the floor slab placed at truck-loading height or railroad car height, approximately 3 to 4 ft above outside grade. This requires placing of compacted fill in the building area for support of the floor slab. This fill generally is placed as one of the first operations in the construction sequence. After it is placed, the foundation excavations are dug and poured, and the walls are constructed.

23.2.2 Settlements

The weight of 4 ft of new fill, representing about 500 lb/ft^2, may cause some settlement to occur over the building area. As a general rule of thumb, settlements are in the range of ½ to 4 in. It is desirable for most of this settlement to occur *before* constructing the foundations and structure.

23.2.3 Retaining Walls

The bottom 4 ft of the perimeter walls must act as retaining walls.

To reduce lateral loads on the walls, it is desirable to tie the floor slab of the building to the perimeter walls. However, it should be remembered that the building foundations and walls will settle differently from the floor slab. Therefore, provisions may need to be built into the floor slab to permit settlement to occur. This might be done by running the reinforcing steel from the perimeter wall into the floor slab, but not pouring concrete in a 2-ft-wide strip around the edge of the floor slab. This 2-ft-wide strip of concrete could be poured later, after building settlements have occurred.

23.3 Slabs in Pile-Supported Structures

23.3.1 Requirements

In areas underlaid by soft soils, pile foundations may be required to support the structure. The floor slab may be supported on grade or on compacted fill. Seldom are such slabs supported on piles.

23.3.2 Grade Beams

It is common practice to place grade beams between pile caps where the soils are soft. The floor slab is above the grade beams.

The weight of the 4-ft-high fill, plus storage loads inside the structure, would cause settlement of the floor slab. However, where grade beams underlie the floor slab, settlement would be prevented. This may cause the floor gradually to take on the appearance of a washboard, with hills over each grade beam and valleys in between. The contractor should be aware that such problems could occur in the future (see Fig. 23.1).

23.3.3 Construction of Floor Slab

To resist lateral loads, it is common for the columns or pile caps to be tied to the floor slab. The floor slab has direct contact with a large area of soil, and therefore provides lateral restraint to the pile caps. This lateral resistance is needed to resist lateral loads due to wind or seismic forces and also to take care of pile eccentricity, particularly in one and two pile groups.

It is common for sliding resistance between the floor slab and soil to be designed for values of 25 to 50 lb/ft^2. These slabs require substantial reinforcing where they are carried into the columns.

23.3.4 Outside Pavements and Underground Utilities

In areas of soft soils, buildings generally are supported on piles. The settlements usually are small. However, the area outside the building may settle, since it is underlaid by soft soils. This places the building in the position of "rising out of the ground." Utility lines outside of the buildings will be pulled down, but their passageways through the building walls will be held in position, requiring frequent maintenance. A better solution is

Fig. 23.1 Large differential settlement.

to provide adequate flexibility in the transition of utility lines from the building to the outside.

Also, pavements outside of the building will gradually settle. The design must be such that this can be accommodated. It may be preferable for sidewalks, loading areas, and other concrete slabs attached to the outside of the building also to be pile supported.

23.4 Buildings over Old Dumps

Dumps have been created in many areas later proposed for use for industrial or commercial buildings. The floor slabs for such buildings can be expected to experience appreciable settlements, on the order of 1 ft or more. The settlements may be due to the weight of fill brought in to grade the surface of the site and also due to slow decomposition of organic materials in the fill.

In addition to causing settlement, the decomposition of organic materials will liberate gas. This gas would tend normally to accumulate under the floor slab of the building and under pavements outside the building. Therefore, a system should be incorporated in the design to collect the gas, and safely vent it to the atmosphere. Such gas (methane) can be dangerous if it is breathed by workers. Also, it is explosive.

24

Landslides

24.1 Types

Landslides occur in slopes composed of all kinds of soil, fill, rock, and mixtures of soil and rock. They may be caused by the following:

Bedding planes.
Planes of weakness in rock or soil.
Undercutting a slope.
Saturation of the ground by rain or water leaks.
Construction of a slope at too steep an angle.
Overloading the top of a slope.
Earthquakes.
Accumulation of rock debris or talus.

Slides may occur as wedges, circular masses of soil, surface or mud flows, or downhill creep of wet soil.

24.1.1 Bedding Plane Slides

Many areas are underlaid by sedimentary rocks with well-defined bedding planes. This sedimentary material may be tilted. In many areas, tilting of 29–40° is common.

An illustration of tilted bedding planes is shown in Fig. 24.1 (also see Fig. 15.4). Figure 24.1 shows a typical landslide along bedding planes.

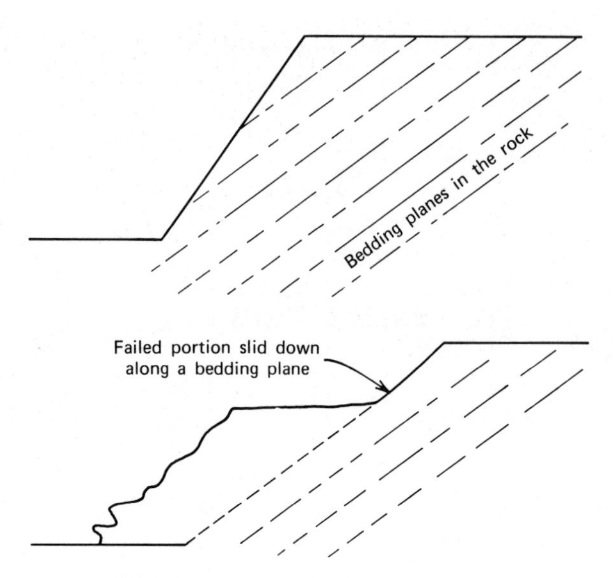

Fig. 24.1 Rockslide along tilted bedding planes.

24.1.2 Fractures and Faults

Many bedrock formations contain fractures or old faults. Faults generally indicate displacement or prior movement, frequently with altered material in the fault zone. A photograph of a fault zone is shown in Fig. 24.2.

Fractures and faults represent weak planes within the rock. Excavation into rocks containing fractures or faults may encounter such a weak plane. If this weakened plane is tilted, the rock sitting above the plane may slide out. Frequently, such slides occur as popouts, or blocks falling out of a slope.

24.1.3 Sand Slides

Sand soil will stand on a slope frequently referred to as the angle of repose. This angle is nearly the same as the angle of internal friction of loose sand, determined by laboratory tests (see Fig. 12.5).

If an excavation is made into clean sand, the walls of the excavation will slough down to the angle of repose. If the sand contains moisture, it may temporarily stand at a steeper angle, rather than slough down to the angle of repose. Small amounts of clay or silt in the sand also may give it greater strength so that the sand can stand at a steeper angle.

Frequently, excavations can be made in clean sand which is temporarily held together by moisture. In such cases, it is common to attempt to hold

Fig. 24.2 Photograph of a fault zone.

the moisture in the sand so that the temporary strength can be retained. This sometimes is done by covering the slope with plastic or other membrane, spraying the slope with sodium silicate or other chemicals, or frequently moistening the slope.

24.1.4 *Clay Slides*

Slides in clay soils generally are characterized by a circular or rotational movement of a large block of material. Such a slide is illustrated in Fig. 24.3. Such slides may be small, or they may involve many acres of land (see Ref. *66*).

24.1.5 *Creep*

The soil on hillsides may be stable during dry weather, but may experience small movements each winter or during heavy rainy seasons. When the soil becomes wet, the soil particles are lubricated. Also, there is a loss of strength. However, a landslide-type failure does not occur. Instead, the soils gradually inch their way down the hillside.

Creep frequently is demonstrated by leaning fence posts, leaning trees, cracks in the soil, and other signs. Some evidence of creep is illustrated in Fig. 24.4.

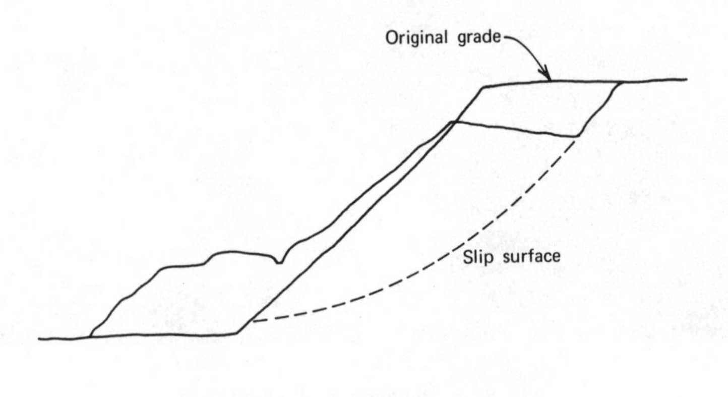

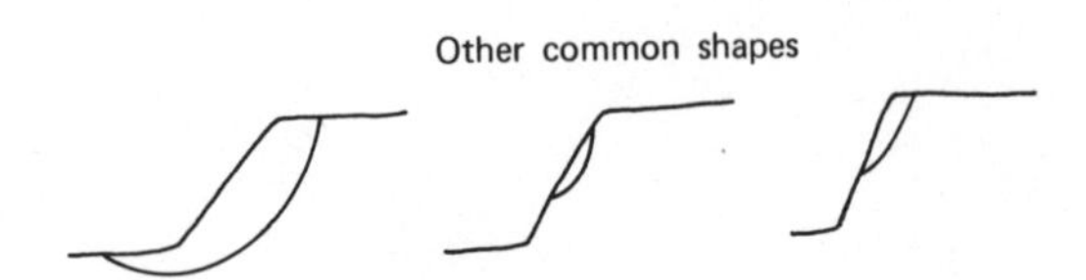

Fig. 24.3 Landslide in clay soil.

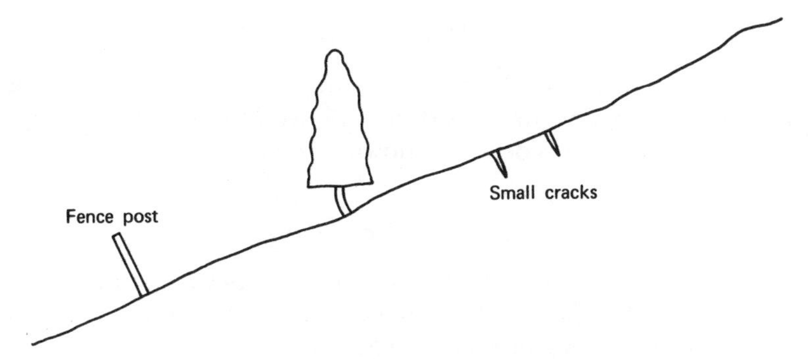

Fig. 24.4 Creep of soil down a hillside.

Fig. 24.5 Mudflow engulfing a residence.

24.1.6 Mudflows

If the soil on a steep hillside becomes sufficiently saturated with water, it may tend to move downhill relatively fast, as a fluid. A photograph of a mudflow engulfing a residence is shown in Fig. 24.5.

24.1.7 Undercutting

Many landslides are caused by undercutting slopes which otherwise would remain stable. This may be done to widen roadways, to extend backyards of houses, or for basement excavations.

Usually city building departments require that undercut slopes be supported by retaining walls or other structures. Temporary undercuts are frequently made to get in footings or other permanent structures. This work should be done in the dry season if possible. Cuts should be opened up in short lengths, or windows, and backfilled before adjacent windows are cut. If the factor of safety of the slope is reasonable for a permanent slope, there should be no difficulty in figuring a reasonable length of window to be cut temporarily for construction. These calculations should be done by the soils engineer.

24.2 Investigation

The size of a potential slide can be estimated to a moderate accuracy. Figure 24.6 shows potential slide surfaces for sand and clay soils. Since sliding could occur on any one of several potential slide surfaces, it is necessary to calculate each one.

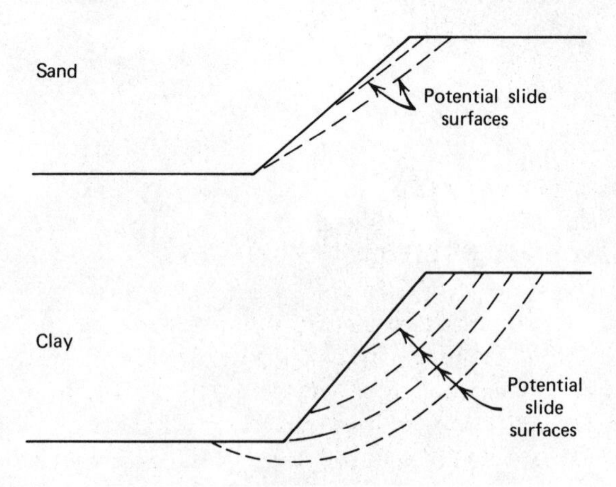

Fig. 24.6 Typical sliding surfaces for sand and clay slopes.

For each assumed sliding surface, a calculation can be made of the force tending to cause sliding. Also, the strength of the soil is figured, based on laboratory tests of soil samples.

Failure of the soil will occur if the sliding force is greater than the strength of the soil. The resisting strength of the soil compared to the sliding force results in a number called the *factor of safety*. The factor of safety equals resisting strength divided by sliding force. For instance, a factor safety of 1.0 indicates that the slope is barely stable. Any small change could result in sliding. A small change which causes sliding is said to "trigger" the slide. If the factor of safety is 2.0, the resisting strength is twice as much as the force trying to cause failure. It is common for permanent slopes to be cut at an angle such that the factor of safety is 1.5.

Before a large cut is made into the soil, it is common to estimate the factor of safety for various proposed angles of excavation. Typical slope angles for various types of soils, and various slope heights are shown in Chapter 21, Section 21.2. These are for permanent slopes. Temporary slopes can be cut steeper (see Chapter 15, Section 15.1.1).

The stability of a slope can be calculated using a method called the "Swedish circle" or method of slices. Such a calculation is shown in Fig. 24.7 (also see Ref. *25*). There are other methods of calculating and shortcut methods for estimating safe slope angles. Many engineering offices have computer programs for performing these calculations.

Sometimes loads are placed at the top of excavations or slopes. These loads may be excavated soil, other material or equipment, or traffic load. It is common to convert these loads into equivalent heights of soil. A slope 20 ft high might be calculated as though it were 26 ft high, or some such number. Then the calculation becomes the same as for any other stability analysis, except that the top load does not have any internal strength (see Chapter 15, Section 15.1.5).

24.2.1 Other Factors

A number of additional factors must be considered. Of primary importance is water. Many slopes are stable until the first sustained rainy season. During wet winter seasons, many cities experience a multitude of landslides. In the winter of 1962, the City of Los Angeles experienced approximately 1700 landslides or mudflows which damaged residential properties. The average cost of damage to each house was $3500.

Measurement wells (also called piezometers) consisting of perforated pipes placed in holes in the ground can be used to observe the water level in a slope. If the water level builds up in the slope, the soil must act as a dam to hold back this hydrostatic pressure. In addition, the water tends to lubricate the soil and also to soften it. These factors act together to encourage a landslide (see Section 24.3.4).

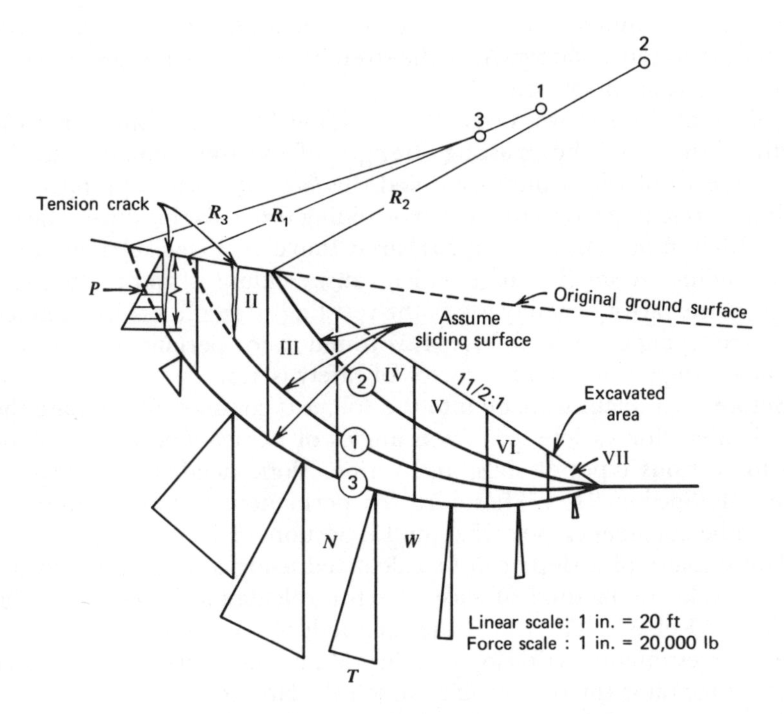

Fig. 24.7 Calculations for slope stability.

CIRCLE 3

Segment No.	*W*	*N*	*T*
I	6,100	3,600	+4,900
II	22,400	17,000	+14,400
III	26,500	23,500	+12,600
IV	23,600	22,800	+6,500
V	17,900	17,900	+1,500
VI	10,000	10,000	−1,100
VII	1,500	1,500	−400
Total		96,500	38,400

$$\text{F.S.} = \frac{\Sigma \text{ N Ton } \phi + \text{CL}}{\Sigma T + ZP}$$

$$\text{F.S.} = \frac{96{,}500 \times .268 + 400 \times 68}{38{,}400 + (0.47 \times 3800)} = 1.3$$

24.2.2 Soil Layers

Usually soils are not a simple, uniform, homogeneous mass throughout the height of a proposed excavation. Instead, they occur in layers. Layers can complicate the investigation of stability of a proposed excavated slope.

Each layer has its own individual strength. Softer layers are a likely zone for a slip plane. In addition, various soils develop their maximum strength at different amounts of deflection. Therefore, a brittle soil combined with a soft soil would not mobilize the total shearing resistance of both soils. The soft soil would develop only part of its strength at the time the brittle soil had reached full stress. The brittle soil would break and lose most of its strength before the soft soil developed its full resisting strength.

24.3 Stabilizing Potential Landslides

24.3.1 Buttresses

Slopes that are potentially hazardous can be stabilized. Some methods include the following:

Compacted buttress fills.
Reinforced-earth buttress fills.
Cribs, consisting of structural frameworks, filled with soil or rock.
Gabion buttresses.
Structural walls.
Structural frames.

Several of these methods of buttressing are illustrated in Fig. 24.8. Usually this work is done in the dry season when the slope is less hazardous.

24.3.2 Removal

Frequently, it is possible to remove an area of potential sliding. This may involve a major change in the site grading plan. As an alternative, the slope area can be dug out, and then reconstructed of compacted fill.

24.3.3 Burial

Occasionally, it is possible to bury a potential landslide. This can be done in special circumstances, for example when there is a steep slope on one side of a gully. By placing a storm drain in the gully, it can be backfilled. The backfill then serves as a buttress to both sides of the gully.

In occasional circumstances, a landslide can be buried by placing a mass of soil on the toe of the slide. This new compacted fill over the lower portion of the unstable slope increases resistance to sliding.

24.3.4 Drains

Water is the enemy of slopes. Therefore, good drainage is essential for any earth or rock slope. Good drainage includes top drainage. Water should be

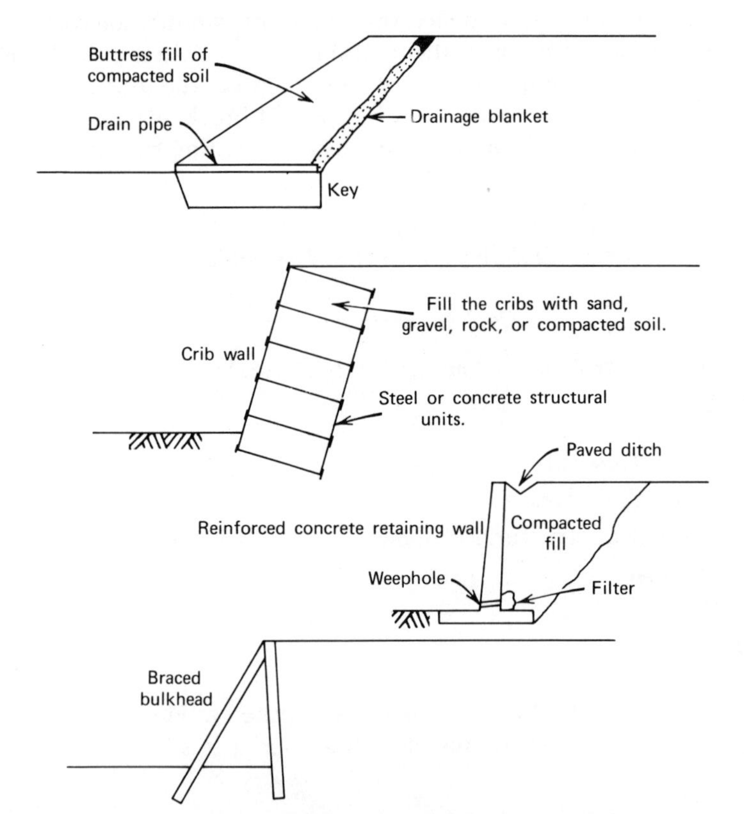

Fig. 24.8 Methods for retaining banks and slopes.

intercepted at the top. The water should be drained back from the slope or around the slope.

An appreciable amount of water can accumulate on the slopes. Therefore, interceptor drains or slope drains should be placed at various intervals on the face of a slope. Such drains are illustrated in Fig. 24.9. The drains can be constructed of asphalt, concrete, or other stable surface. They need to be cleaned out after each heavy rain and before the start of each rainy season.

Interior drains may be required to prevent the buildup of water within a soil mass. If the slope is being constructed, drainage devices can be built into the slope. However, if a cut slope or other existing slope must be drained, drains must be inserted into the slope. One method commonly used is to drill horizontal holes back into the slope and to case the holes with perforated pipe. Such an installation is shown in Fig. 24.9. Usually, such holes are 3 to 6 in. in diameter and lined with plastic or metal pipe.

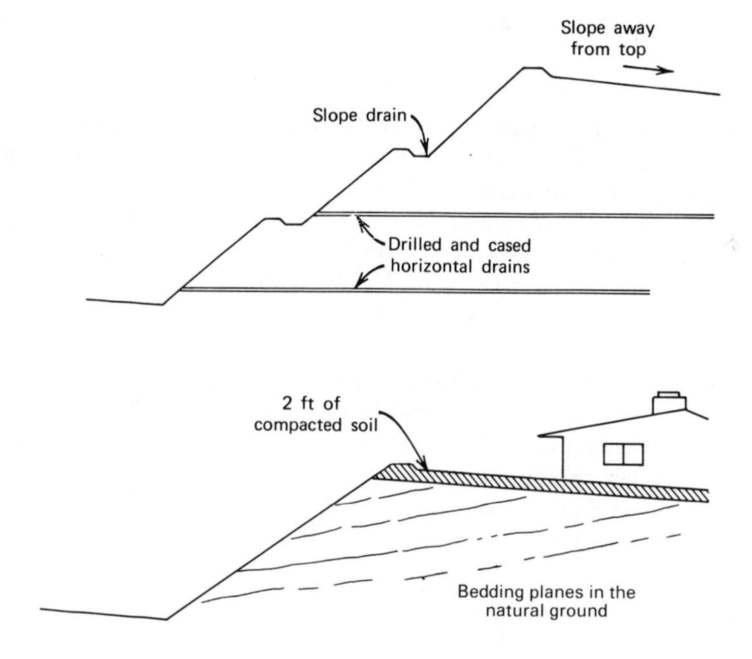

Fig. 24.9 Slope drainage methods.

Such horizontal drains usually slope upward approximately 5° to flow freely. Such drains frequently are installed to depths of 100–300 ft back into the slope. The drains are spaced horizontally at intervals on the order of 25 to 50 ft, center to center. Grading to provide a flat surface may have sliced off topsoil and exposed inclined bedding planes in a rock formation. If these bedding planes are tilted, water may get in from sprinkling or from rainfall. In such cases, it is desirable to cover the bedrock. A layer of compacted soil can act as such a cover (see Fig. 24.9).

Erosion protection for the surface of slopes should be placed as early as possible during construction.

Finally, avoid cesspools, leaching fields, or other sources of water on all sloping or hillside sites.

24.3.5 Other Methods

Other methods of correction may include lowering the height of the slope or flattening the slope angle by giving up some flat land at the top of the level area. Also, some small slides have been stabilized by pins through the slide zone and also by injection of chemicals into the slide zone (see Ref. *67*).

24.4 Active Landslides

Frequently, it is important to develop temporary protection when some signs of sliding start to develop. Usually, this occurs in a heavy rain. Large sheets of building plastic can be used to cover a slope temporarily. Sandbags can be placed on the plastic to keep it from blowing away. Number 2 reinforcing bars 3 ft long can be stabbed through the sandbags, plastic sheeting, and into the ground to help hold the plastic in place.

Water running over the top of a slope should be diverted away with temporary curbs of sandbags.

Any cracks or fissures developing in the ground should be filled with silty or clayey soil or covered over with plastic sheeting or asphalt.

After the rains stop, the conditions should be investigated by a soils engineer. His work may include borings, laboratory tests, and calculations. Also, he may set special casings in the borings to find the slip surface, if slippage is occurring (see Ref. *38*, p. 1042, and Fig. 14.6).

25

Retaining Structures

25.1 Types

Retaining structures may come in many sizes and styles, according to the designer's ingenuity. Some retaining structures are well designed and functional. Others may have the appearance of strength but may be more a showpiece. Retaining structures may also be used in combinations.

Some common types are as follows:

Vertical concrete cantilever walls.
Vertical concrete buttress walls.
Soldier piles with lagging, and frames of vertical and battered piles.
Soldier piles, freestanding.
Soldier piles with tiebacks, sheet piles with tiebacks, or reinforced earth.
Rock gravity walls.
Buttresses of compacted soil.
Bin-type wall or gabion wall.

The various types of walls listed above are illustrated in Fig. 25.1 (also see Chapter 16).

25.2 Retaining Structures

Retaining structures usually are constructed first, and the backfill is placed between the wall and the excavated slope.

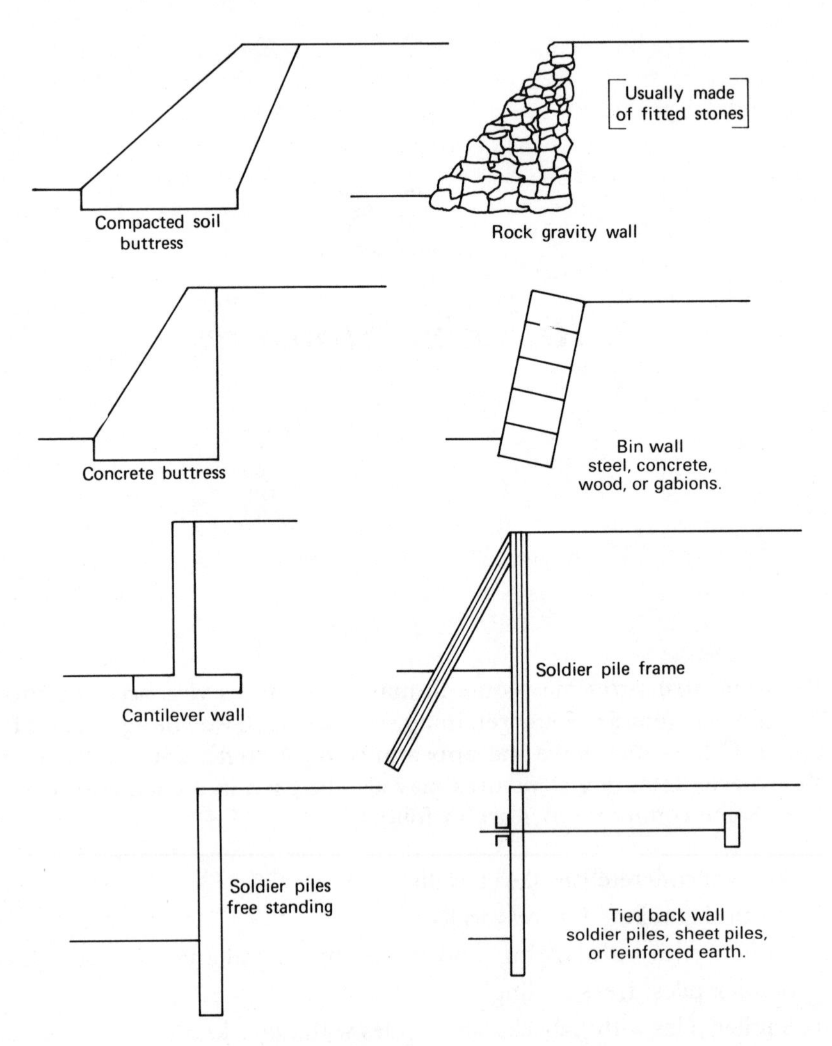

Fig. 25.1 Examples of retaining structures.

The backfill soil should be selected from the available on-site materials. If the on-site materials are wet soils or expansive clays, they may be unsuitable for use as backfill; it may be necessary to import suitable soil for backfill.

The backfill soil will impose a lateral pressure on the retaining structure. The amount of lateral earth pressure will depend on many factors, such as the type of soil, the degree of compaction, dry or wet backfill, and whether the wall is flexible or rigidly braced.

When backfill is placed behind a wall, it is assumed that the wall can bend laterally a small amount. The block of soil behind the wall starts to deflect. This mobilizes the shear strength of the soil. The strength of the soil plus the strength of the wall are enough to prevent the block from sliding. This soil strength thereby reduces the lateral pressure on the wall. This frequently is called the active soil pressure. This concept is illustrated in Fig. 25.2.

For the soil to deflect, the wall must move. The amount of movement required to develop the soil strength for the active earth pressure generally is ⅒ to 1% of the height of the wall. It is common for a wall 10 ft high to deflect ½ in. or more at the top of the wall, and for walls 20 ft high to deflect 1 in. or more.

Sand generally is considered good backfill soil. The active pressure is relatively small, and the required deflection of the wall is small. Also, sand drains rapidly if rainwater or other water gets into the backfill.

By contrast, loose dumped clay soils, particularly if saturated, can be very different. The wet soils creep, and therefore will not maintain their strength. Therefore, the active pressure gradually becomes larger. Such a backfill may impose 300% of the load imposed by sand.

If the clay is placed at proper moisture content as a compacted fill, it usually develops substantial strength. In this case, the lateral active pressure is reasonably low, and the clay is a satisfactory material.

Some clay soils are expansive. They swell when they become saturated. Such soils would be extremely undesirable for backfill, since they would swell when they got wet and impose very large lateral pressures on the wall.

Occasionally, an excavation can be cut neat to the outside wall dimensions. In this case, a form can be placed for the inside of the wall and the concrete can be poured directly against the soil. Since no backfill is placed, the lateral pressure on the wall theoretically is zero. In such cases, it is general practice to design for a lateral earth pressure in the range of 25 to 30 lb/ft^2 times the height of the wall. This number is sometimes called the equivalent "fluid pressure;" the pressure would be the same as for a fluid weighing 30 lb/ft^3.

The lateral pressure behind walls probably is larger than generally is assumed in the design of walls. In compacting backfills behind walls, the soil must be at a pressure greater than the active pressure. However, it is believed that a compacted soil becomes relatively rigid. Therefore, a small deflection of the wall would develop the active pressure within the soil.

Also, it has generally been found that walls designed and constructed for lateral pressures of 30 to 35 lb/ft^2 have performed satisfactorily. This satisfactory record may be due to the factor of safety in the steel and concrete of the wall. The exceptions have been due to unplanned pressures, such as water saturation.

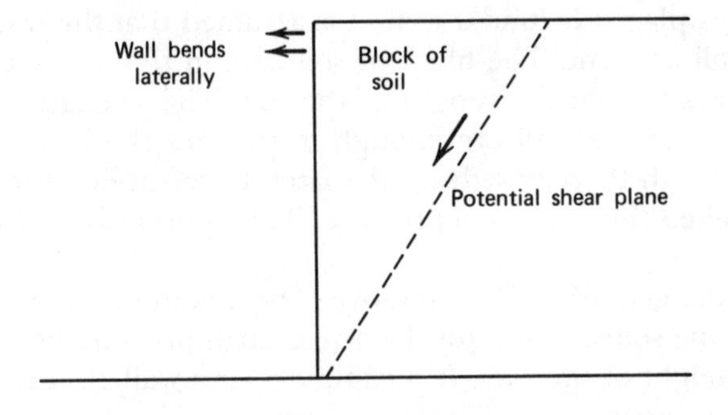

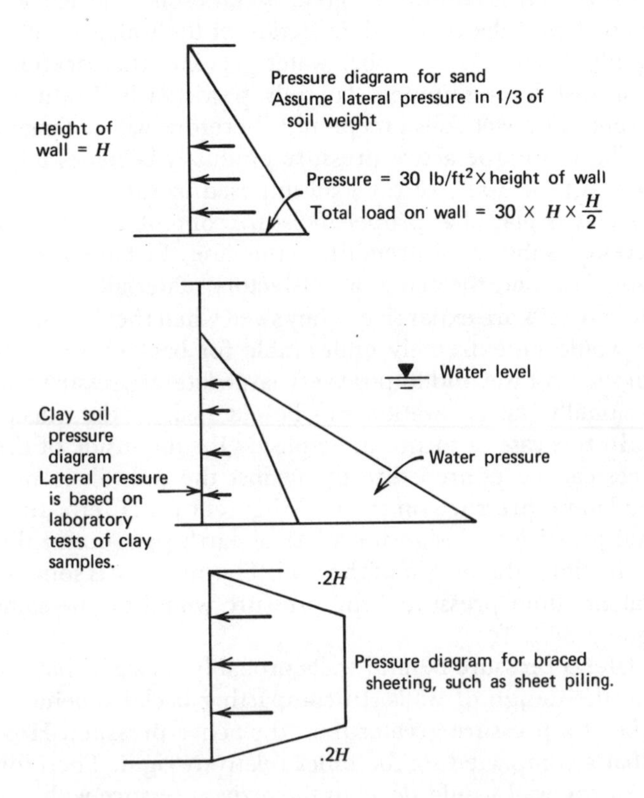

Fig. 25.2 Active soil pressure on retaining structures.

The vertical height at which soil will stand can be estimated from the following formula:

$$H = \frac{2C}{W}$$

where C = cohesion or shearing strength of clay soils.
W = weight of the soil in pounds per cubic foot.

Therefore, for a soil with a strength of 500 lb/ft^2 and a weight of 100 lb/ft^3, the maximum height to which a vertical bank would stand is approximately 2 times 500 over 100 equals 10 ft. This calculated height should be reduced to provide some factor of safety.

25.3 Drainage

Most retaining structures are provided with weep holes or other drainage facilities to prevent the backfill soils from becoming saturated. A typical weep hole is illustrated in Fig. 25.3. Weep holes usually are placed at intervals of about 10 ft center-to-center along walls. Weep holes are a nuisance to build, and contractors have trouble doing them right. It is necessary to provide free flow to and through the weep holes but to prevent clogging and loss of backfill through the holes. Practice was to place a sack filled with gravel opposite each hole back of the wall, but soil backfill works into the gravel and clogs it. Prevention of clogging requires a graded filter providing layers of successively coarser grain size, from that of the soil backfill to that of the gravel at the weep hole entrance. This is difficult and costly, but there are commercially available mixtures of sand and rock which can be placed in a large bag at a weephole entrance. The sand keeps soil from clogging the gravel. Just at the back end of the weep hole, the sand will wash out, but it will leave the gravel and thus form a graded filter.

Now that filter cloths are commonly available these problems are much more easily treated. If the gravel to be placed at the weep hole entrance is encased in filter cloth having the right size openings to prevent passage of the backfill soil, the necessary filtering is simply provided.

Where lawn watering, heavy rains, or other large sources of water drain into the backfill behind a retaining wall, it is necessary to increase weep hole and drainage design. One means for this is to run a gravel blanket the full height of the wall (see Fig. 25.3). A layer of filter cloth between the gravel blanket and backfill provides needed filtering and prevents clogging of the gravel. Without adequate drainage, the continued saturation of the fill will cause greater lateral pressures on the wall and may weaken the foundation soils and result in greater tilting of the wall.

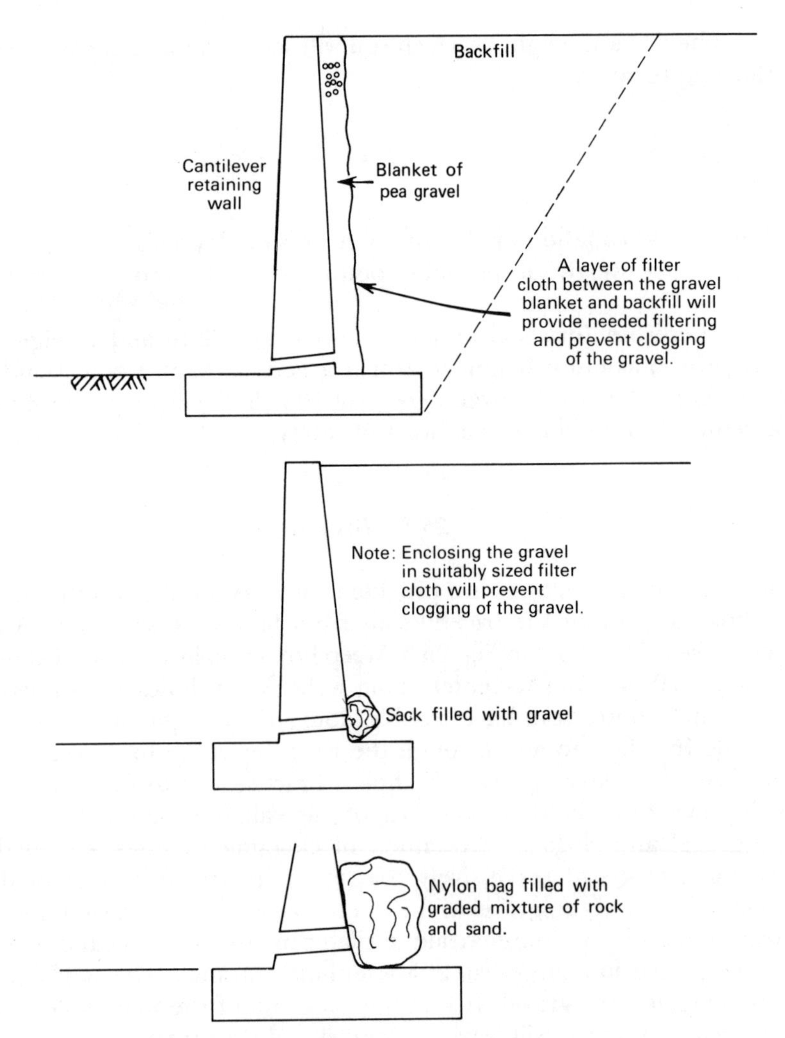

Fig. 25.3 Retaining wall drainage systems.

25.4 Attachment to Other Structures

Retaining walls must be expected to move or tilt slightly during and after construction.

If such structures are doweled or otherwise fastened to buildings, the lateral deflections will cause cracking at the intersection of the wall with the adjacent building.

At right angle external corners of walls, vertical cracks occur frequently.

Retaining walls should be designed so that they can tilt slightly; ¼ to ½ in. is common. Joints may be cut in the wall to permit this flexibility.

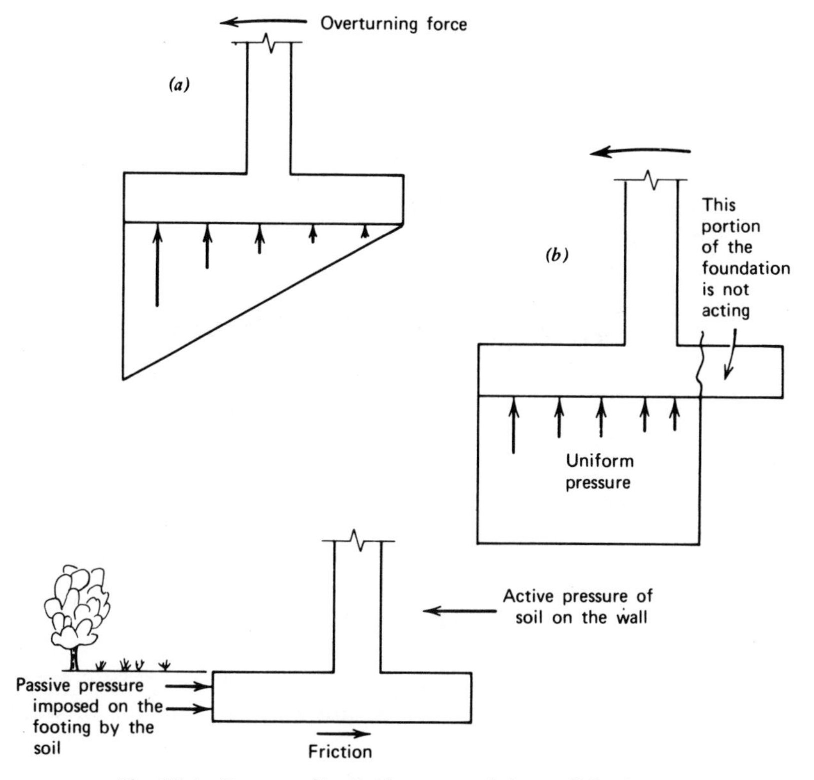

Fig. 25.4 Pressure distribution on retaining wall footings.

The joints may be filled with joint compound and can be backed up with metal or permanent flashing-type material.

25.5 Foundations

The foundations of cantilever walls will have two primary forces—overturning and sliding, as shown in Fig. 25.4.

Overturning will cause pressures at one edge of the footing to be much larger. This edge pressure cannot exceed the soil bearing value. Sometimes the bearing pressure is assumed to act over only a part of the footing area, as shown in Fig. 25.4*b*. The back part of the footing is considered to be not acting.

Sliding or skidding of the footing is resisted by "passive resistance" in front of the footing. Typical values for passive resistance are in the range of 200 to 500 lb/ft^2 times depth in feet. In addition, there is sliding resistance on the bottom of the footing. This resistance usually is in the range of 100 to 500 lb/ft^2.

26

Linings and Membranes

26.1 Uses

Reservoirs, canals, storage ponds, evaporation ponds, artifical lakes, oil storage reservoirs, and ditches may be lined to prevent loss of water or other fluid.

26.2 Soil Linings

The most common lining material is soil. Soil linings generally are referred to as clay linings. Such linings may be constructed of clay, or mixtures of sand, silt, and clay.

Bentonite, a very plastic clay, is mined in several parts of the United States, and is sold in sacks or in bulk. Bentonite frequently is used for lining. Usually it is mixed with the native soil.

In areas where clay is not easily available, reasonably good linings have been constructed of mixtures of available soil. On one reservoir, a mixture of fine sand and silt was used. With the proper moisture content and good compaction, a relatively tight lining resulted.

Soil liners usually are 2 to 3 ft thick and are placed in layers 6 in. thick. Each layer is compacted with larger rollers. On slopes which are 2 horizontal to 1 vertical, or flatter, the soil layers can be placed conforming to the slope. Compaction usually is performed with sheepsfoot rollers, or other compaction equipment, operating up and down the slope. Usually the liners consist of at least three layers, and more commonly are four layers thick. The soils should be at optimum moisture content, or slightly wetter.

Soils compacted slightly on the wet side tend to have a lower permeability. Also, they are more plastic and do not tend to develop vertical cracking during the compaction process.

Deformations in completed reservoirs lined with clay indicate that soil can accommodate some movement without cracking. This is a very desirable asset. It is difficult to compact the soil on steeper slopes. This could be done by placing the soil in horizontal layers the width of the rolling equipment and then trimming off the outside 2 or 3 ft of soil after compaction is completed.

Compacted fills usually are compacted to densities of at least 90% of the maximum density according to the modified AASHO method of testing. It is desirable to keep the soil lining wet. If it is allowed to dry out, it may shrink and crack, becoming less watertight.

26.3 Concrete Linings

Concrete linings are used in many projects. Usually the concrete is cast in place. Sometimes it is placed as shotcrete (gunite). These linings usually are not reinforced, but the shotcrete lining may contain a minimum amount of mesh, primarily for control of cracking due to shrinkage. Sometimes, however, it may be desirable to reinforce the concrete lining where it is on steep slopes, or where hydrostatic back pressure may develop behind the lining.

The placing of gunite concrete or poured-in-place concrete involves the use of low-water-content concrete, particularly in slope lining. Low water content also is necessary to limit shrinking and cracking of the concrete, which would result in leaking. Concrete linings usually are 4 in. thick. Expansion joints usually are placed 20 to 30 ft, center-to-center. Where canals are lined, they usually are lined with a machine especially designed for placing and finishing the concrete.

Concrete linings encounter difficulties in several circumstances, as follows:

Collapsing soil. In many arid areas where water canals are constructed, there are "streams" of low-density soil crossing the path of the canal. Generally these low-density soils can be recognized only by testing. Settlements of 1 ft occur frequently, and sometimes settlements are several feet where a canal crosses such soil.

Expansive soil. Expansion of adobe or gumbo-type soils can cause cracking of a concrete lining, particularly at transitions with structures.

Loose backfill. Pipes, culverts, and other structures may cross under a canal or reservoir lining. When the reservoir has been full for some time, it must be expected that some water will leak through the lining and get into the subgrade soil. The backfill may not be as stiff as the natural soil and may settle, cracking the concrete and causing more leakage.

26.4 Commercial Linings and Membranes

Many commercial lining materials are manufactured. These include many types of plastic. Sheets of polyvinyl chloride are available in various thicknesses and at reasonable cost. Some common plastics are reinforced with fiberglass or other materials to increase their strength. Butyl rubber and other materials are also commercially available as linings. Ref. *68* and *69* cover applications of linings.

Linings also are constructed in sheets or panels. These panels may be constructed of asphalt, asphalt-soaked fibers, or precast concrete.

Plastic sheets require a carefully prepared and smooth bed. Rocks and other sharp objects are not permitted in the supporting bed. Frequently sand is used to prepare a suitable bed.

The plastic sheets usually come in rolls and can be unrolled and cut into panels in place. Where each joint occurs, the panels usually are overlapped, and can be joined by heating irons, which weld the sheets together. Usually, the joints are overlapped, and two passes are made with the irons to result in double bonding of sheets.

Because the plastic sheets are sensitive to weather and deterioration by sunlight, and are also subject to holes, from being poked, it is common to cover the plastic with a layer of sand, gravel, or other protective soil. Usually the soil thickness is 6 to 12 in. The spreading operation is done by hand, with some mechanical assistance, to prevent damage to the plastic.

Rolled asphaltic concrete linings generally are 3 to 6 in. thick. They are placed and rolled following procedures common to roadway paving. Usually the asphalt is spread with mechanical equipment, and two layers of asphalt are placed. The top layer is placed and rolled in lanes perpendicular to those in the underlying layer.

Concrete sometimes is placed by special forming techniques. One such technique is the "fabriform" sold by Intrusion-Prepakt (Contech). This is a nylon cloth form placed in the desired location on a slope or channel bottom. It can be placed on a dry surface or underwater. When the form is in the desired location, concrete is pumped into it, causing it to expand and resemble a mattress. This method is indicated in Fig. 26.1 (see Ref. *70*).

26.5 Permeable Protective Linings

In some cases, permeable linings are desirable because they provide stability to slopes and resistance to erosion, and yet free drainage to avoid buildup

Fig. 26.1 "Fabriform" concrete canal lining as furnished by Intrusion-Prepakt (Contech).

of hydrostatic back pressures. Gravel is used frequently, as well as manufactured porous concrete planks and similar devices.

26.6 Underdrains

Some reservoirs or canals must be emptied or the water lowered considerably in short periods of time. This is called rapid drawdown. In spite of all care in the use of materials and good construction techniques, it must be expected that linings will leak to some degree. Therefore, the soil embankment behind the lining must be assumed to be saturated.

If the embankment walls are fairly steep, on rapid drawdown the water in the soil tries to flow back into the reservoir. This water pressure pushes against the back side of the concrete lining and causes the concrete slab to lift or deflect slightly. This type of behavior is somehow progressive: failure may not occur during one emptying, but with several occurrences, the slab can be successively deflected until it breaks.

Some soils may be sufficiently strong to remain stable during rapid drawdown of the reservoir water level. Also, concrete liners sometimes can be made sufficiently strong to resist hydrostatic uplift. However, in many cases it is necessary to relieve the water pressure in the slope. This is done by constructing "underdrains" in the embankments, or relief holes with back-flow valves in the lining.

Underdrains may consist of (*a*) blankets of sand, (*b*) trenches at intervals of 10 or 15 ft on centers filled with sand, (*c*) porous concrete or porous wall pipe, and (*d*) perforated or slotted pipe, surrounded by sand, filter cloth, or suitable filter material. These drains must be designed as filters, so that soil cannot wash into the drains and clog them. The design of filters (graded filters) or selection of filter cloth having the proper size openings requires a knowledge of the grain sizes of the soil in the embankment. Such designs should be made by an engineer experienced in this work (see Ref. *21*, p. 66).

Damage of concrete lining slabs is due to deflection of the slab. Deflection is caused by the sustained flow of water into the thin space between the embankment and the concrete slab. If the concrete slab is sufficiently heavy, the water pressure in back of the slab is not sufficient to raise the slab. Also, if the rate of emptying the reservoir is so slow that water will drain out of the embankment, the slab will not be uplifted.

A photograph of slab uplifting is shown in Fig. 26.2. Generally, such movements are more likely to occur on steep slopes. Flatter slopes involve

Fig. 26.2 Uplifting of concrete lining slab.

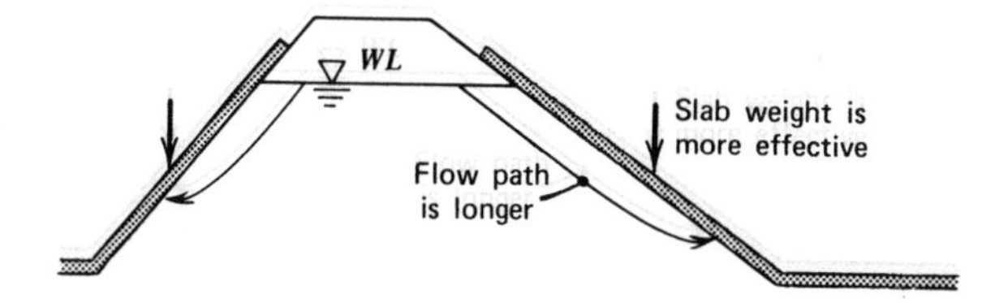

Fig. 26.3 Flow path at drainage under canal lining.

longer flow paths, and the weight of concrete is more effective in resisting uplift (see Fig. 26.3).

The worst conditions for uplift are as follows:

1. Steep slopes, such as 1 to 1.
2. Silty soils.
3. Rapid drawdown.

26.7 Blowouts

Blowouts are very similar to rapid drawdown. When a reservoir is emptied quickly, the hydrostatic pressures under the bottom slab may be sufficient to cause the slab to rise and to break the slab or cause a blowout. Generally, this condition is relieved by placing a drainage blanket or other drainage collection devices under the bottom slab. They may connect to a drainpipe leading away from the reservoir. As an alternative, there may be pop-up valves, relief wells, or other devices that permit the temporary excess water pressure under the slab to drain upward through the slab.

26.8 Maintenance

Liners generally require periodic maintenance. Original construction should be done anticipating that future repairs and maintenance will be necessary. Large washouts occur from time to time. Usually, they can be traced back to sustained leakage through the lining, which caused softening, then erosion of the soil.

27

Potential Damage

27.1 General

Construction operations frequently cause damage to adjacent property. It is the duty of the contractor to work in a careful manner, planning each phase to avoid damage to adjacent property. It is also the duty of the adjacent property owner to notify the contractor of any indication that damage is starting, and that additional protective measures should be taken. The adjacent property owner must take all steps necessary to protect that property from damage, even though he or she intends to make a claim against the contractor.

The most dangerous phases of construction include demolishing old structures, excavating for basements, blasting rock, site dewatering, setting and stressing anchors or braces to shore up the sides of the excavation, and driving pile foundations. During these phases, damage may occur to adjacent structures, or to underground utilities, streets, or sidewalks.

After construction is completed, there is the potential for damage to the completed work. Such damage may include high water levels, with wet floors and seepage into basements; foundation settlements; cracked floor slabs; corrosion; vibration due to machinery operation; and slope erosion.

27.2 Excavation

A general discussion of excavation and bracing is presented in Chapters 15 and 16. Although excavations may stand up satisfactorily and the bracing system may prevent cave-ins, frequently the adjacent streets settle and crack;

utility lines settle, break, and flood out excavations, or cut off service in a neighborhood; and the foundations and sidewalks of adjacent structures stretch, settle, and crack. Bracing generally is required for vertical or steep excavations exceeding depths of 6 ft. Sometimes deeper vertical excavations are permitted where the soils are proven to be sufficiently strong to stand vertically. Building codes usually specify shoring and bracing, but give no guidelines. State industrial safety codes may give some specific requirements.

The design of shoring and bracing systems is almost always left to the contractor. The contractor may have his or her own forces do the design or may hire an engineer to design a system. The contractor may pass on to a specialty subcontractor the responsibilities to design and install the shoring system.

Occasionally, damage is due to serious underdesign of the shoring system and actual failure during excavation. Much more often however, the bracing and shoring system remains intact. However, the shoring deforms and yields small amounts as excavation proceeds. Another possibility is that the job superintendent wants the excavation to proceed rapidly and may not take the time to be sure all of the bracing system is in place and completed in all respects before making the next stage of the excavation.

To limit the amount of lateral yielding of the excavations, the bracing system must be preloaded or prestressed as it is constructed; otherwise, the bracing system must deform in order to develop its design stress, thus permitting some lateral yielding of the excavation wall.

In addition, shoring systems frequently are based on the active pressure of the soil. The active pressure develops only after the soil has deformed. Sufficient deformation to develop active pressure may be on the order of 1% of the depth of the excavation. Therefore, for an excavation 20 ft deep, properly shored and braced, measurements on many projects have shown that lateral movements of the top of the ground may be 2 to 3 in. For soft soils the movements may be greater, and for firm soils they may be less. However, a movement of 2 to 3 in. laterally can be enough to substantially damage utilities and adjacent structures.

If the excavation is 40 ft deep, lateral movements range from 4 to 6 in., with much greater resulting damage to adjacent structures.

The greatest stretching of the ground is close to the edge of the excavation, but stretching may extend back from the excavation a distance equal to the depth of the excavation, or even greater distances, such as twice the depth of the excavation. Therefore, for an excavation 40 ft deep, buildings or utilities within 40 ft can be expected to stretch and develop some cracks. On softer soils, buildings 80 ft away may be affected. In sandy soils, the movements usually are much smaller and much less likely to be a problem, except in saturated sands, which can be difficult to work with and prevent loss of ground.

In design, the usual procedure is to make the shoring system as light and low-cost as possible. The lateral soil pressures are assumed to develop into the fully active case, resulting in the least possible lateral pressure on the shoring system. As mentioned previously, substantial soil deformation is required to develop the fully active case. Therefore, lateral deformation of the soil must be expected.

It may be possible to reduce lateral deformation by using a stiffer, stronger bracing system. Such a design can be developed assuming that the fully active case does not develop. Lateral pressures then would more likely be somewhere between "at rest" and "active." The soils engineer should be asked whether it is safe to use the active pressure and how much deformation may result.

In making surveys of adjacent buildings, it is good to have some guide as to when to become concerned with the measured settlements. In most buildings, column spacings are 20 to 25 ft on centers. As a rough guide, it is desirable to limit differential settlement between columns to the range of ¼ to ½ in. If differential settlement reaches 1 in., some cracking will become apparent. At 2 in. it is expected that cracking will be serious, and there may be structural damage. Therefore, when survey readings show that a column has dropped even as little as 1⁄16 in., some action should be taken. If settlements progress to as much as 1 in. with respect to the adjacent column, it is time to take definite corrective action.

27.3 Methods of Detection

It is important that the contractor inspect regularly the street and adjacent buildings for indications of movement. One of the most common and best systems is a well-planned grid of survey points. Elevation monuments should be established at several locations in each adjacent building, and on the street and sidewalks. The monuments should be surveyed regularly. Two or more reference bench marks should be established at a distance far enough away from the site to be free of disturbance from excavation and construction activities.

Where deep excavations must be dug or where the site is underlaid by soft soils, it may be desirable to take horizontal measurements between elevation monuments, and also to set up separate survey points to measure lateral stretching of the ground. Frequently lateral movement is detected by observing new cracks, and measuring cracks on a regular basis. However, survey points measured regularly by tape can provide better information.

Also, vertical holes can be drilled outside of the perimeter of the proposed excavation, and casings can be set in these holes to measure lateral movements of the ground as the excavation proceeds (see Chapter 15, Section 15.1.6). Special casing is installed in such holes, and a precise inclinometer device can be used to record gradual deformations of the ground near the excavation.

In addition to measurements, good quality photographs before and during construction can be very helpful to indicate changes, which may occur so slowly that the job superintendent is not aware of them except by referring back to previous photographs.

27.4 Additions to Existing Buildings

Frequently new additions are constructed which tie into an existing building. The old building has experienced its settlement; however, the new building will settle during construction, and perhaps for some time after construction. The amount of cracking at the joint between the old building and new building can be reduced if some separation is left between the two buildings as long as possible before they are joined.

Permanent embankments of soil several feet thick may be placed for industrial warehouses with the floor slab at dock loading height. Such fills weigh several hundred pounds per square foot and can be expected to settle perhaps several inches. Therefore, the embankments should be placed as early as possible, long before foundations are placed in or near these embankments. Settlement readings can be taken to chart the progress of settlement of the embankments.

It may be desirable to stockpile soil on the site for backfilling purposes. Frequently the stockpile is placed at the edge of the site, adjacent to an existing structure. The weight of the stockpile may be expected to cause some settlement of the ground and may drag down the foundations of the adjacent building.

27.5 Site Dewatering during Construction

Substantially lowering the normal groundwater level in the construction area can have a profound influence on adjacent structures.

In many cases, adjacent buildings are supported on untreated wood piles. As long as the wood piles are saturated completely, they will not rot or deteriorate. However, if they are dried out, they are susceptible to rapid rotting and decay. Several cases of serious damage to pile foundations of existing buildings have been due to temporary lowering of the groundwater level.

If the lowering of groundwater levels lasts only for a short period of time, and then the groundwater conditions are established at their previous level, there may not be time for any substantial damage to the piles. Generally, serious damage occurs when dewatering is extended over many months or years.

In some cases, the groundwater level under adjacent structures has been kept high by recirculating water back to the structures and recharging it into the ground (see Refs. *41* and *42*). The design of injection wells for this purpose requires the services of someone experienced in this field.

Lowering the groundwater level for a long period of time also causes consolidation and settlement of the soil. When the water level is high, the soil particles are buoyed up by the water, and in effect are lighter. When the groundwater level drops, these soil particles become heavier. Therefore, if the water level is dropped 10 ft, the soil weight is increased roughly 400 lb/ft^2. If the water level is dropped 20 ft, the effective increase in soil weight is 800 lb/ft^2. This increase in weight can cause compressible layers of soil to settle several inches. The zone affected may extend several hundred feet away from the edge of the dewatered area.

The most common method of dewatering is sump pumping. Usually, one or more pits or sumps are dug in or adjacent to the excavation, extending a few feet below the bottom of the general site excavation. Rim trenches lead water to the sump. A submersible or suction pump is used to pump water from the sump. Occasionally, an excavation is dug as deep as possible before installation of any dewatering system. The groundwater level tends to flow up into the bottom of the excavation, creating a soft condition, and in some cases, even a quick condition. In one such case the writers are familiar with, the bottom became so unstable the contractor could not operate equipment on the bottom. The contractor claimed that this "quicksand" condition made it impossible to support the building on spread foundations, and that piles would be required. Fortunately, the contractor was persuaded to put in a drainage system; within three days the bottom was stable and the contractor then proceeded to form and pour the foundations.

In some cases, sump pumping is not sufficient to adequately dewater a site. Either the bottom continues to stay soft, or boils form, causing loss of ground in the subgrade. In these cases, a deeper dewatering system is required. The most common system is well points. Well points are shallow wells and are usually installed around the perimeter of the site. The well point system can pull the groundwater level down approximately 15 ft. If deeper dewatering is required, it is necessary to put a second stage of well points at a lower elevation.

Another system for deep dewatering is to install several wells around the perimeter of the site. Wells permit dewatering to any depth desired.

It is very important in any kind of dewatering system that the water pumped out runs clean. If soil is pumped out with the water, it indicates removal of material, which could lead to the formation of cavities or areas of soft subgrade. The contractor should check the discharge frequently and take samples of water in large glass jars. The jars should be allowed to stand to see if any sediment settles out or if soil particles can be observed in the water.

27.6 *Driving of Pile Foundations*

The act of driving piles can cause problems, such as heaving of the ground surface (see Chapter 20, Section 20.8). Pile driving also can cause lateral movement of the soil. This is a great hazard when driving into cohesive soils adjacent to a deep excavation. If the soils in the area are loose sands, the vibration may result in some compaction of the soil, causing settlement of nearby buildings.

Not only buildings, but utility lines buried in the ground also are susceptible to damage due to vibration, ground heaving, lateral displacements, or settlement.

Frequently it becomes necessary to drive piles in such locations. Various methods are used to accomplish pile driving without creating damage. Some such methods are as follows:

1. Pilot-jet a hole to be followed by a pile and pile driving.
2. Jet the pile down, with a jet system which goes down with the pile.
3. Predrill a hole at the location of each pile. Usually, predrilled holes are slightly smaller in diameter than the pile (see Chapter 20, Section 20.9).
4. Use a pipe pile, driven open-end, and clean out the pile as driving progresses.
5. Use a heavier, slower acting hammer with more cushion blocks. Heavy, slow-acting hammers tend to develop less shock and somewhat more push on the pile.

27.7 *Shock and Vibration*

Large pile driving rigs put a large amount of shock energy into the ground each time the hammer strikes the top of the pile. These shock forces travel rapidly through the ground to adjacent structures, and often can be felt by people in the adjacent buildings. Vibration of the building due to the shock forces can in some cases be damaging to adjacent structures. This is not very common. Vibrations strong enough to cause structural damage generally are very upsetting to people. Therefore, complaints from the neighbors usually are sufficient to stop pile driving operations before structural damage actually occurs.

Very often, a recording seismograph is used to measure ground motion vibrations during pile driving, blasting, or other construction operations causing shock impacts on the soil. This includes operation of pavement breakers, large pumps or compressors, and heavy truck traffic.

A seismograph is shown in Fig. 27.1. Generally, accelerations or displacements are measured on the ground surface, or on floor slabs in the

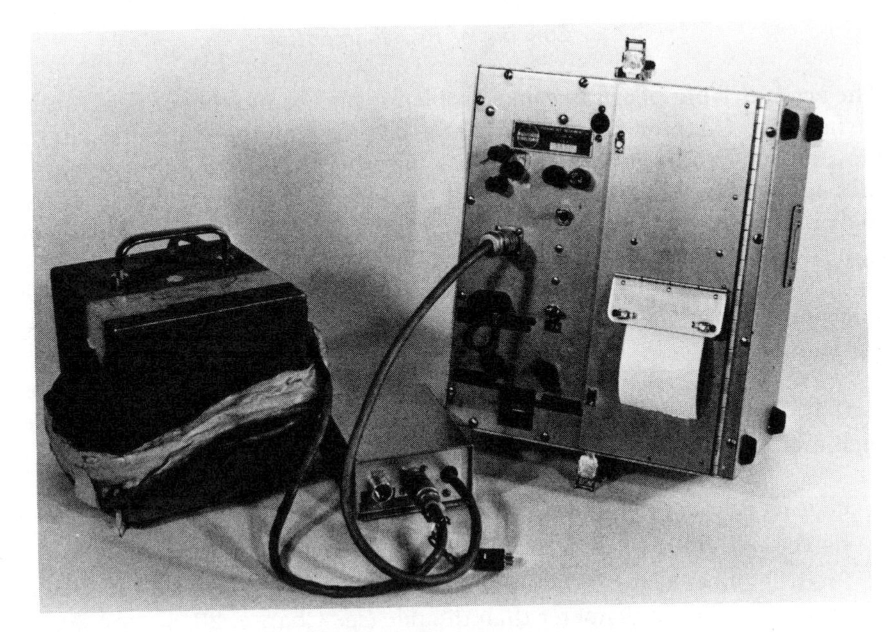

Fig. 27.1 Recording seismograph.

adjacent buildings. The measurements are made of vertical vibrations, and horizontal vibrations in two directions. The ground motion is amplified and printed on paper or film. These records are reviewed by someone experienced in this work, and the results are used to reach opinions regarding possible damage to adjacent structures.

Figure 27.2 shows a plot of amplitude of vibration measured at three locations away from the point of the blast. A series of curves indicating the effect of vibrations on people and structures are also shown in Fig. 27.2. Assuming that blasting is being done 360 ft from an adjacent building, the amplitude of motion is 0.0065 in. at a frequency of 11 Hz. This point is shown plotted in Fig. 27.2. This indicates that the level of vibration is severe to persons but safe with regard to the structure. Since human beings are very sensitive to vibrations, readings of this kind are very important to show that construction operations can proceed without structural damage to adjacent structures. Such data frequently are used in court (see Ref. *71*).

27.8 Damage after Construction

Water leaks through basement walls or floors can be due to holes in the waterproofing system. Also, many basements are designed with a drainage system to relieve the pressure around the basement walls. Drainage systems are difficult to build and to prevent from clogging. Therefore, the drainage system should be constructed in such a way that maintenance is not too difficult. If the system does not work, the contractor will usually be called back to correct the problem and get the system into a working condition. Frequently, soil will wash through the gravel pack around drainpipes and settle out inside the drainpipe. Therefore, drainpipes must be laid out so that they are easy to clean out.

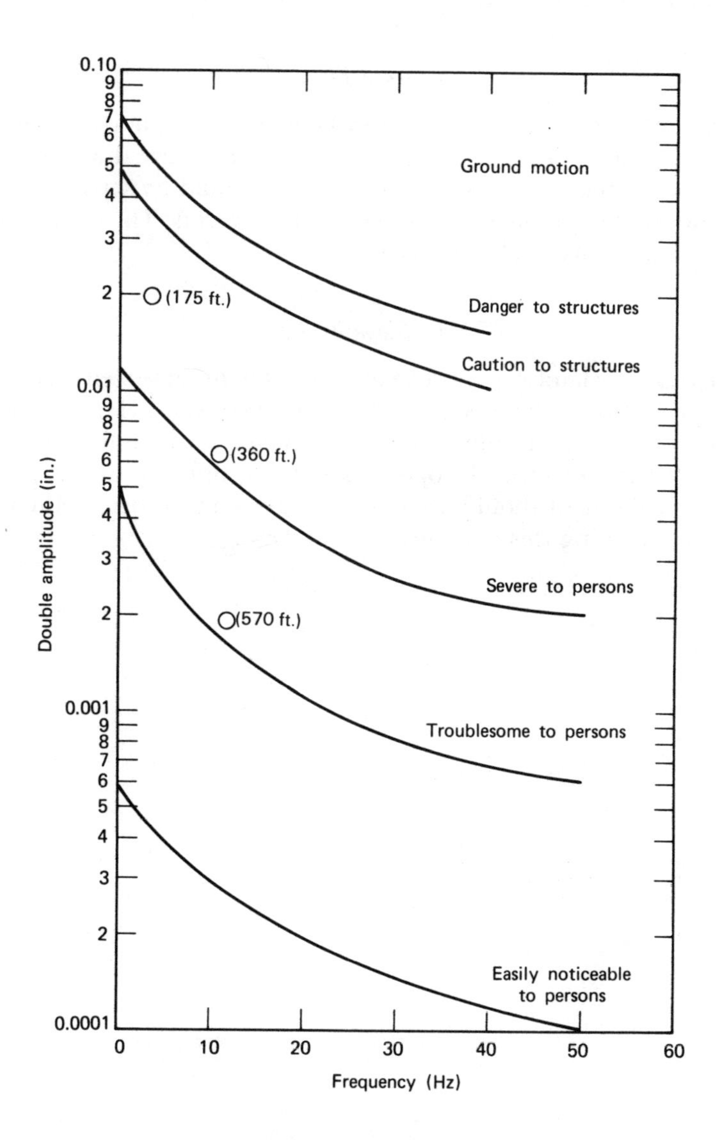

Fig. 27.2 The amplitude and frequency of vibrations with consequent effect on structures and persons.

27.8.1 Settlements

A new building frequently will experience some settlement during construction and after completion. This settlement may cause some downdrag of the foundations of adjacent structures. In the bidding stage, make sure that this possibility has been adequately considered by the structural engineer and soils engineer. Damage due to this cause should not become the responsibility of the contractor.

27.8.2 Cracked Floor Slabs

Floor slab settlements and cracking can be a result of poorly compacted backfills in trenches cut for utility lines, or occasionally are due to swelling and heaving of subgrade soils. Generally, slabs should be separated from columns and walls to permit some differential movements. The cracks should be filled with joint material, such as Thiokol.

27.8.3 Slope Erosion

In many cases, new landscaping vegetation has not become well established when the project is ready to be turned over to the owner. Heavy rains can cause erosion of slopes, resulting in disagreements between the contractor and owner. Repair of eroded slopes is a tedious, difficult, and expensive process. The contractor should make sure to have some understanding with the owner concerning this type of potential damage.

28

Grouting and Chemical Injection

28.1 General Applications

Several kinds of grout and chemicals for chemical stabilization of soil are commonly used in the construction industry.

Grouting may be done to fill voids in the soil, or to fill voids or cracks in rocks. Also, grout curtains act as a barrier for the flow of water. Some grouts are installed to displace earth and to compact it in place.

Chemical grouts generally flow into a coarser-grained sand or gravel soil, and are used to cement the soil together. Chemical grouts also may be used to stop the flow of water through the ground.

28.2 Cement Grout

Cement grout may be a mixture of cement and water, or sand, cement, and water.

On many construction projects, it is necessary to fill voids in the ground. Soft soil or peat, which is compressible, must be stabilized to limit future consolidation and settlement. This type of intrusion grouting is sometimes described as consolidation grouting, and is briefly discussed herein. Grouting for curtain walls and to reduce seepage around dams, which is a highly specialized grouting technique, is not discussed at length in this book. Several companies have the equipment and trained personnel for this work. Two projects indicating the usefulness of this method are as follows:

The site of a proposed structure in Florida was found to be underlaid by large cavities formed by the solution of limestone bedrock. These cavities were estimated to be at least 30,000 cu yd in volume. The solution cavities started about 50 ft below ground surface and extended to a depth of as much as 100 ft. Borings indicated that the top few feet of the cavities were filled with water, and below that the cavities contained extremely soft soil of low density in the range of 50 to 60 lb/ft^3. The drill stem on a drill rig dropped through the soil with no resistance. The initial grouting into the cavity was sand and water. The sand was pumped in through pipes extending down into the cavities. Also, relief pipes were installed to permit excess water to drain out. Over 3000 yd of sand were placed in the cavities, followed by 700 yd of a grout mixture of sand, cement, and water. Later borings showed that the cavities had been filled and that there was a substantial increase in the strength of the soft soils in the cavities. Soil density changed from a range of 50–60 lb/ft^3 to 70–90 lb/ft^3. The purpose of filling was primarily to prevent future sinkhole-type collapse of the overlying soil which had been "bridging over" the cavities.

On a second project, a building was constructed over an area which had once been a peat bog. The foundations extended down to firm soil. The floor slab was supported directly on an earthfill dumped on top of the peat bog material. Settlements of the floor slab varied, and were as much as 1 ft. A mixture of sand, cement, and water was pumped into the peat bog material, starting at the bottom of the soft soil. By pumping up the bottom, and then gradually raising the pipes and pumping at higher elevations, the grout was forced into the peat bog material, tending to compress and compact it. Approximately 2000 yd of material were pumped into the peat bog material, representing an addition of approximately 45 lb of material to 1 ft^3 of the peat bog soil. On completion, the floor slab was lifted substantially. It should be noted, however, that it is frequently necessary to return and grout a second time when dealing with soft, compressible clay or peat soils.

28.3 Soil-Cement Displacement Grouting

The injection of soil-cement to displace and compact soils has become a relatively important part of soil stabilization. It is done most frequently to repair buildings damaged by settlement. This work is done by specialty subcontractors.

If the soils underlying the site are at relatively low density, settlements of footings or floor slabs may be substantial. If the soils are compared to the compacted soil density, the resulting degree of compaction of the existing soils may be only 60 to 70%.

The characteristics resulting from low density are as follows:

1. Low bearing value.
2. Settlement under load.

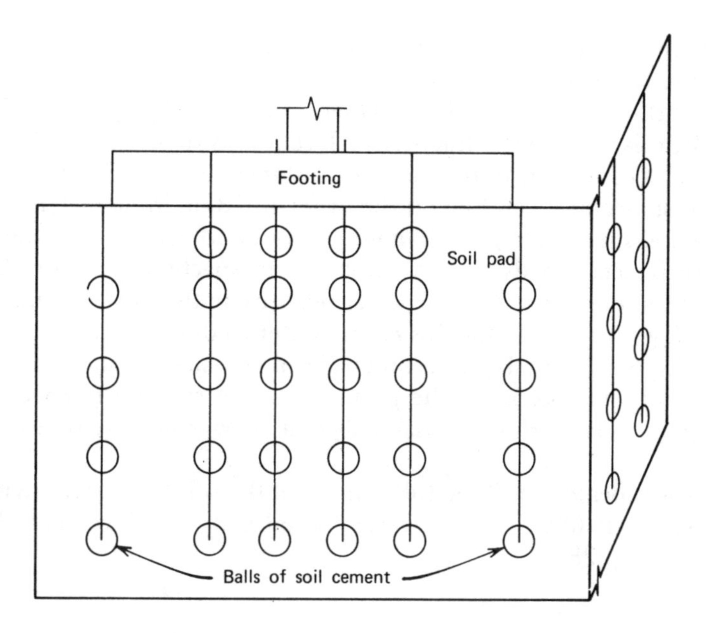

Fig. 28.1 Sketch of pattern of intrusion of balls of soil-cement.

3. Additional settlement with substantial increase in moisture content.
4. Additional settlement if the soil is vibrated or subjected to shock loading, such as by an earthquake.
5. The potential for liquefaction and soil instability in cases of extreme earthquake shocks applied to low-density saturated sand.

Assume that a building is supported on footings resting on low-density soil, and many of the footings have settled. The possible solutions include: (*a*) underpinning each footing, one at a time, with new and wider footings or pile foundations, or (*b*) increasing the strength of the underlying supporting soil, and raising the old footings.

Usually, increasing the strength of the soil is most desirable because it causes the least interference with operation of the structure.

Compaction can be accomplished by injecting material into the ground. This forces the soil to compress sideways and vertically. The soil then occupies less space and therefore is more compact.

The material must be injected in such a way that the ground surface will not be moved or heave upward. Foundations must be raised slowly, in several steps, and in unison. The material injected is a low moisture content, nonplastic, soil-cement. This material usually is installed at a number of select locations. This material does not "flow." It is confined to the place injected. It must be controllable.

A pump at the ground surface takes a specific volume of soil-cement and pumps it down a grout pipe to a specific depth below the ground surface. At this point, a ball of soil cement is formed. Because it is dry and nonplastic, this ball stays at the desired location. A definite known volume of material is contained in the ball. For instance, the ball might contain 2 or 3 ft^3 of soil-cement, and be on the order of 18 in. in diameter.

The soil surrounding the ball is squeezed and compacts. If the soil is sand, it compacts quickly. If the soil is silt, compaction is slower. After a ball of soil-cement is inserted, the water pressure in the soil pores increases. A time delay is necessary for this excess water to drain away.

During the time delay, a second ball can be placed at another location through another pipe. Later the grout pipes are raised a few feet above the previous balls, and additional soil-cement balls are inserted into the earth.

This is repeated until the full volume of the soil "pad" underlying the footing has been filled with properly spaced balls of soil-cement. This is indicated in Fig. 28.1.

28.4 Planned Compaction

Laboratory tests can be performed to estimate the degree of compaction required to achieve a desired bearing capacity of the soil. Samples of the natural soils are compacted at various higher densities, and shear tests are performed to measure the new strength. A graph of a series of such tests is shown in Fig. 28.2. This graph indicates the shear strength of the soil at its natural density of 70% relative compaction. The graph also indicates the increase in strength at higher densities.

As an example, assume that a bearing value of 4000 lb/ft^2 exists, and it is desired to raise it to 6000. The bearing value graph of Fig. 28.2 indicates that approximately 83% compaction is required in order to obtain the desired strength.

In compacting the soil from 70 to 83%, there would be a loss in the volume of the soil. Assume that the soil mat to be compacted is 20 by 20 ft in plan dimensions and 20 ft thick. This is about 300 yd^3. This soil when recompacted would occupy 85% of the total previous volume, or about 255 yd^3. This would require adding 45 yd^3 to the volume of the soil mat, which would be done with soil-cement. This amounts to about 18 lb of soil-cement added to each cubic foot of soil. Obviously, the numbers given here are for illustration purposes only.

Levels must be set up, and ground surface elevations must be read frequently during injection to make sure that lifting of the footings is done in small steps.

The soil-cement usually is 7 or 8% cement and 92 or 93% soil. The moisture content is kept as low as possible.

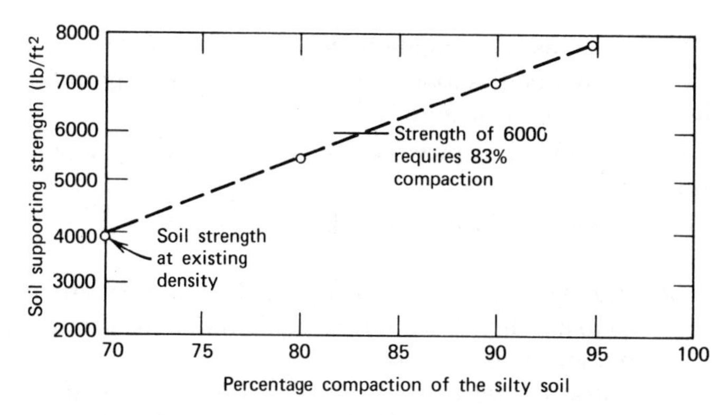

Fig. 28.2 Graph of shear strength increase with compaction of soil.

The soil is a graded sand, with some fines, such that the void spaces between larger particles are filled with appropriately sized smaller particles. The subcontractor should select the material to be used.

Grouting, mud jacking, and pressure injection have been used for a number of years. In many cases the results have been good, although occasionally they were not. Probably the most important consideration is that skill and specific experience in this work are essential to doing a good job.

28.5 Chemical Grouting

Chemical grouting generally is done in sandy soils or soils that are permeable. The chemicals are in a liquid form, frequently with a consistency not much thicker than water. The chemicals must be very fluid, because they must flow through the soil. The purpose of this kind of grouting is not to displace soil. Instead, these chemicals flow around the soil grains. These chemicals harden or set up and cling to the soil grains. They tend to cement the soil together. Some chemicals after setup are hard. Others are like firm gelatin. Also, the chemicals fill the small void spaces between grains of soil and make the soil less pervious. Water cutoffs can be developed with chemical grouts.

The primary problem is deciding whether the soil will "take" the chemicals. Sand usually is suitable as long as it does not contain too much fine material. Some silt in the sand generally is acceptable. However, it is generally impossible to stabilize silt and clay soils by chemical grouting. Occasionally, chemicals can penetrate moderately well into fairly silty soils but some laboratory testing is necessary to determine this.

Commonly used chemicals include the following:

1. *Sodium Silicate.* This was originally developed as the "Joosten" process. There are several modifications of this process, operating under a number of names, including Siroc and Hunts.
2. *Chrome Lignun.* This is a process that uses byproducts from the paper pulp industry.
3. *AM9.* This material is like a stiff gelatin when it sets.

Generally, chemicals are mixed in large tanks, and then are fed through a manifold system to a pump which forces the fluid into the ground. Pumping pressures are frequently on the order of 20 lb/in.^{2}.

Specialty contractors work with these chemicals. Considerable experience is needed for successful operation. Since hardening of the chemicals is based on a chemical reaction, it is possible that the chemical characteristics of the soils, groundwater, or even the mixing water can have an adverse effect on the reaction of the chemicals in the ground. The reaction time must be delayed so that the chemicals can penetrate the soils before starting to set up.

The resulting strength of stabilized sand is generally good for silicate-type grout (Joosten process). Generally, compressive strengths are greater than 100 lb/in.^{2}. Frequently they are 300 to 400 lb/in.^{2}.

This process is used for underpinning of foundations, stabilizing slopes of excavations, and stabilizing the soils above the arch in tunneling through sandy soil. In addition, the chemical can be sprayed on slopes to reduce erosion from wind or rainfall.

Inspection of grouting operations is very difficult. By probing, it is possible to outline the extent of cemented material fairly well. It is difficult to get core samples of stabilized soil for laboratory testing. Instead, laboratory samples are prepared using predetermined amounts of the chemical.

28.6 Soil Stabilization for Pavement Base Course

Since crushed rock or gravel sometimes is very expensive, a substitute is to upgrade available soil. Cement is frequently used to add strength to on-site soil. The amount of cement used varies with the type of soil, and with the strength desired. However, the amount of cement usually is 5 to 10% of the soil used. The resulting strength usually is in the range of 300 to 800 lb/in.^{2}. The soil-cement mixture is compacted at optimum moisture, just like any fill soil (see Ref. 72).

Frequently lime is used to improve the characteristics of clay soils. Clay soils usually are a poor and unstable subgrade to support the base course for pavements. They are plastic and have a high plastic index (see Chapter 12, Section 12.4). The addition of lime reduces the plastic index, and the strength of the compacted clay may be doubled. Lime also tends to dry out the soil if the clay is too wet for compaction.

29

Legal Aspects of Construction Operation

29.1 General

A number of attorneys specialize in the construction field exclusively. Many others become involved to a considerable extent in construction work. Why? At least partly because underground work involves some unknown conditions. Never can an engineer have perfect knowledge of the soil, rock, and water conditions underlying a site. The owner of a large construction firm once said, "We never lost any money above ground."

The legal aspects of soil and foundation conditions on construction operations are discussed in much greater detail in Ref. *73*. Therefore, only a few comments are made in this book.

29.2 Available Information

Many contractors feel that it is not their responsibility to know anything about the soil conditions at a site. They believe this is a problem of the owner and of the architect or design engineer. By contrast, many owners feel that it is not their responsibility to know anything about the soil conditions at the site, and that this information must be developed and used by the contractor. Some owners are reluctant to submit soils data to the contractor, even if submitted on the basis of "information only."

Therefore, obtaining sufficient data on a site sometimes is a problem. In general, however, the trend is more and more toward the owner obtaining information and making it available to the contractor. Therefore, both become informed concerning the site conditions, and both recognize that there are limits to the amount of information available. This probably is better than some existing conditions in which contractors prefer to bid "blind" on soil conditions, and assume that all other contractors will bid "blind." All such bids then contain contingencies concerning the unknown soil conditions. In a way, providing data to the contractor requires that he or she make more specific estimates of construction procedures based on these data. This takes time. Contractors do not like to spend this amount of time preparing bids. They can make the more specific estimates if they get the job.

29.3 Changed Conditions

Usually, contractors have only 30 days to prepare a bid for a job. Therefore, they rely to a considerable extent on information given to them.

Some changes will occur on almost all projects. Usually if there are truly greatly different conditions than anticipated, contractors are able to get awards for extras. After all, the project would have been designed differently if the different condition had been known from the beginning. However, the owner should be told ahead of time that he or she *may* have to pay more if the conditions do change.

Generally, the law does not expect the drawings to be perfect. However, the law does require good practice and reasonable care on the part of the designer. Also, the designer or owner is expected to act reasonably when something unexpected does show up, requiring a change in the drawings.

For a contractor to claim "changed" or latent conditions, he is required to demonstrate that he correctly evaluated the soil conditions as represented, and that he visited the site to study the terrain and learn what he could from nearby outcrops, cuts, or excavations.

For additional information see Refs. *73* through *76*.

30

Construction Specifications

30.1 Purpose

Construction specifications add to and supplement the contract drawings. They should spell out the quality of the materials to be used, the work which the contractor is to perform, the time schedule for the contract, the basis for compensation, and the procedure for authorizing changes to the contract price.

Construction specifications, as a rule, leave the contractor free to select methods of doing the work, including excavation, design and installation of temporary shoring or bracing, foundation construction, and backfilling.

The specifications spell out the responsibilities of the contractor for performance and for providing protection to the workers and the public. The owner, engineers, and others are to be protected against claims arising from the contractor's acts (negligent or otherwise) and property damage.

Provision is made for evaluation of claims for delays, authorized changes, and claims for extra work. Several bases may be used, such as unit prices, agreed-upon prices, time and material, and—in the event of no agreement—arbitration. Many specifications nominate the engineer as the arbiter of disputes and the interpreter of the intent of the contract documents.

Alternative forms of construction considered by the engineers as potentially economical are included frequently on the drawings and in the specifications. However, specifications seldom make provision for alternates proposed by contractors.

Information concerning soil, water, and rock conditions may be included on the drawings, in other contract documents, or in the specifications.

Where a foundation includes caissons or piles, specifications set forth in detail the basis for pricing that part of the work. The types of piles that may be used, the lengths and resistance to which they shall be driven, permissible tolerances, type of hammer, class of concrete (if required), methods of depositing concrete, and criteria for acceptance of the driven piles are presented. Also, if test piles or pile load tests are required, the test program is described.

The specifications should tell the contractor what he is expected to do, the responsibilities he will assume, and the services that the owner will provide. The specifications also try to protect the owner from the contractor's acts and from unreasonable claims.

30.2 Disclosure of Information

It is the professional (and legal) obligation of the engineers to furnish contractors with all the information they have concerning site and soil conditions. Since this information is the property of the owner, the owner must direct the engineer to give all such information to the contractor.

With reference to the soil information, this means that all information available in field notes and boring logs must be included. Notes which report, for example, "Lost all water at 26 ft," must not be passed over. Furthermore, if the engineer has reclassified the soil samples to better describe the soils, that fact should be recorded. In addition, the samples and field logs must be preserved and available for inspection by the contractors.

In many areas of the country, certain general conditions exist which are known to local engineers and contractors, but not necessarily to others. Such latent conditions might be expansive surface soils; caliche or hardpan; an artesian head of water extending several feet above the street level; the probability of encountering gas; two water levels as the result of perched water above the true groundwater; or buried shallow peat deposits or old swamps that have been filled over. This kind of information is easily overlooked in writing specifications.

In the case of buildings adjacent to a proposed excavation, all information concerning the depths and dimensions of the building foundations must be furnished. Otherwise, the absence of such information should be plainly stated.

The contractors, however, must make every reasonable investigation themselves of site and soil conditions, and be able to prove that they did investigate them.

Some specifications state that soils information and reports may be inspected at the office of the engineer or the owner. Frequently jobs are bid and won by contractors who did not come in to look at this information.

The technical sections of specifications may require that the contractor review soil mechanics literature to determine the significance of the soil data presented. For example, where dewatering of a site or compaction of soil is necessary, the percentage of fines in the soil or the shape of the grain size curves helps to assess the feasibility of the work. Also, the specifications should be compared with the appropriate sections of the local building code (where applicable).

The contractor may feel that some provision of the specifications cannot be accomplished without extraordinary methods and expense. He should explain his concern to the engineer and request an interpretation. A more difficult alternative (if he gets the contract) is to demonstrate in the field the difficulties of meeting the specified requirement, and to secure a modification.

The last resort in the face of an impossible requirement is to submit a qualified bid. If the bid is low and the engineers can agree legally to consider it, a change may be negotiated.

30.3 Shortcomings in Specifications

It is an unfortunate fact that specification writing is an unpopular chore in most design offices. Some specifications seen by the writers were so inappropriate that one wondered if the wrong book was sent along with the plans. Usually the specifications are prepared by borrowing paragraphs or even chapters from other specifications—prepared years before for other jobs. With scissors and Scotch tape a specification for the job at hand is pasted together. Therefore, many specifications contain contradictions and vague wording, and sometimes important features are not included.

References

1. T. William Lambe, "The Structure of Compacted Clay," *Journal of the Soil Mechanics and Foundations Division*, Proceedings Paper 1654, American Society of Civil Engineers, New York, May 1958.
2. Raymond N. Yong and Benno P. Warkentin, *Introduction to Soil Behavior*, Macmillan, New York, 1966.
3. *Earth Manual*, U.S. Department of Interior, Bureau of Reclamation, Washington, D.C.
4. *Standard Test Method for Classification of Soils for Engineering Purposes*, ASTM D2487, American Society for Testing and Materials, Philadelphia, Penn. Also see Refs. 13 and 3.
5. *Standard Test Method for Liquid Limit, Plastic Limit, and Plasticity Index of Soils*, ASTM D4318, American Society for Testing and Materials, Philadelphia, Penn.
6. *Standard Test Method for Shrinkage Factors of Soils*, ASTM D427, American Society for Testing and Materials, Philadelphia, Penn.
7. *Standard Practice for Description and Identification of Soils (Visual-Manual Procedure)*, ASTM D2488, American Society for Testing and Materials, Philadelphia, Penn.
8. C. C. Ladd, "Swelling of Compacted Clay," S. M. Thesis, Massachusetts Institute of Technology, Cambridge, Mass., 1956.
9. *Handbook of Ocean Engineering*, McGraw-Hill, New York, 1949.
10. *Standard Method for Penetration Test and Split-Barrel Sampling of Soils*, ASTM D1586, American Society for Testing and Materials, Philadelphia, Penn.
11. Gordon Fletcher, "The Standard Penetration Test—Its Uses and Abuses," *Journal of the Soil Mechanics and Foundations Division, Paper 4935*, American Society of Civil Engineers, New York, July 1965.
12. *Standard Practice for Thin-Walled Tube Sampling of Soils*, ASTM D1587, American Society for Testing and Materials, Philadelphia, Penn.
13. K. Terzaghi and R. B. Peck, *Soil Mechanics in Engineering Practice*, 2nd ed., Wiley, New York, 1967.

14. T. R. Dames, "Practical Shear Tests for Foundation Design," *Civil Engineering*, December 1940.
15. RQD stands for rock quality designation and is described in the following publications: (*a*) Don U. Deere, "Technical Description of Rock Cores for Engineering Purposes," *Rock Mechanics and Engineering Geology* **1**(1), 17 (1964). (*b*) *Standard Test Method for Determining Deformability and Strength of Weak Rock by an In Situ Uniaxial Compressive Test*, ASTM D4555, American Society for Testing and Materials, Philadelphia, Penn.
16. *Standard Method for Field Measurement of Soil Resistivity Using the Weiner Four-Electrode Method*, ASTM G57, American Society for Testing and Materials, Philadelphia, Penn.
17. J. J. Jakosky, *Exploration Geophysics*, Trija, Los Angeles, Calif., 1950.
18. Schultz and Cleaves, *Geology in Engineering*, Wiley, New York, 1955. Also Robert Crimmins, Reuben Samuels, and B. Monahan, *Construction Rock Work Guide*, Wiley, New York, 1972.
19. *Standard Specification for Wire-Cloth Sieves for Testing Purposes*, ASTM E11, American Society for Testing and Materials, Philadelphia, Penn.
20. *Standard Method for Particle-Size Analysis of Soils*, ASTM D422, American Society for Testing and Materials, Philadelphia, Penn.
21. R. B. Peck, W. E. Hanson, and T. H. Thornburn, *Foundation Engineering*, Wiley, New York, 1953.
22. *Standard Method for Direct Shear Test of Soils under Consolidated Drained Conditions*, ASTM D3080, American Society for Testing and Materials, Philadelphia, Penn.
23. *Standard Test Method for Unconfined Compressive Strength of Cohesive Soil*, ASTM D2166, American Society for Testing and Materials, Philadelphia, Penn.
24. *Standard Test Method for Unconsolidated, Undrained Compressive Strength of Cohesive Soils in Triaxial Compression*, ASTM D2850, American Society for Testing and Materials, Philadelphia, Penn.
25. D. Taylor, *Soil Mechanics*, Wiley, New York, 1948.
26. H. R. Cedeargren, *Seepage, Drainage and Flow Nets*, Wiley, New York, 1967.
27. *Standard Test Methods for Moisture-Density Relations of Soils and Soil-Aggregate Mixtures Using 5.5-lb. Rammer and 12-in. Drop*, ASTM D698, American Society for Testing and Materials, Philadelphia, Penn.
28. *Standard Test Methods for Moisture-Density Relations of Soils and Soil-Aggregate Mixtures Using 10-lb. Rammer and 18-in. Drop*, ASTM D1557, American Society for Testing and Materials, Philadelphia, Penn.
29. *Special Procedures for Testing Soil and Rock for Engineering Purposes, Special Test Publication 479*, Committee D-18, American Society for Testing and Materials, Philadelphia, Penn., 1980.
30. *Standard Practice for Classification of Soils and Soil-Aggregate Mixtures for Highway Construction Purposes*, ASTM D3282, American Society for Testing and Materials, Philadelphia, Penn.
31. *Standard Test Method for Classification of Soils for Engineering Purposes*, ASTM D2487, American Society for Testing and Materials, Philadelphia, Penn.
32. *Standard Test Method for Bearing Capacity of Soil for Static Load on Spread Footings*, ASTM D1194, American Society for Testing and Materials, Philadelphia, Penn.
33. *Standard Method for Non-Repetitive Static Plate Load Tests of Soils and Flexible Pavement Components for Use in Evaluation and Design of Airport and Highway Pavements*,

ASTM D1196, American Society for Testing and Materials, Philadelphia, Penn.
34. *General Catalog*, Soil Test, Inc., Evanston, Ill.
35. *Standard Method of Testing Piles under Static Axial Compressive Load*, ASTM D1143, American Society for Testing and Materials, Philadelphia, Penn.
36. *Uniform Building Code*, International Conference of Building Officials, Whittier, Calif., Chapt. 29.
37. Wayne C. Teng, *Foundation Design*, Prentice-Hall, Englewood Cliffs, N.J., 1962.
38. G. Leonards, *Foundation Engineering*, McGraw-Hill, New York, 1962.
39. Aerospray 52 Binder is a commercially sold spray-on protection for slopes and faces of excavations. It is sold by American Cyanamid Company, Bound Brook, N.J.
40. R. L. Peurifoy, *Construction Planning, Equipment and Methods*, 2nd ed., McGraw-Hill, New York, 1970.
41. J. Patrick Powers, *Construction Dewatering: A Guide to Theory and Practice*, Wiley, New York, 1981.
42. James D. Parsons, "Foundation Installation Requiring Recharging of Ground Water," *ASCE Construction Journal*, September 1959.
43. Weldon S. Booth, "Tiebacks in Soil," *Civil Engineering*, September 1966.
44. Norman Liver, "Tension Piles and Diagonal Tiebacks," *ASCE Construction Journal*, July 1969.
45. Charles I. Mansur and M. M. Alizadeh, *Journal of Soil Mechanics and Foundations Division, ASCE*, March 1970, p. 495.
46. "References on Slurry Trench Excavation," *Western Construction Magazine*, October 1968.
47. Manufacturers of narrow vibrating rollers are Essick Manufacturing Company, Los Angeles, Calif.; Vibro plus Products, Inc.; Stanhope, N.J., and others.
48. *Handbook of Drainage and Construction Products, ARMCO Manual*, Middletown, Ohio. Also, *Handbook of Steel Drainage and Highway Construction Products*, American Iron and Steel Institute, New York, various dates.
49. Merlin G. Spangler, *Soil Engineering*, International Textbook Company, Scranton, Penn., 1969.
50. "Compacted Earth Fill for a Power Plant Foundation," *Civil Engineering*, August 1961.
51. "The Distribution of Sulphates in Clay Soils and Groundwater," The Institution of Civil Engineers, Paper 5883, London, England, 1953.
52. Jacob Feld, *Construction Failure*, Wiley, New York, 1968.
53. J. McWhorter and J. Burrage, *Construction Contract Law*, in preparation.
54. William W. Moore, "Experiences with Predetermining Pile Lengths," *ASCE Transactions*, 1949.
55. Robert D. Chellis, *Pile Foundations*, McGraw-Hill, New York, 1961.
56. J. D. Parsons, "Difficulties in the New York Area," *Proceedings ASCE Journal of the Soil Mechanics and Foundation Division* **92**, 43–64. Nai C. Yang, "Relaxation of Piles in Sand and Organic Silt," *Journal of the Soil Mechanics and Foundation Division, ASCE*, **95** (SM21), *Proceedings Paper 8123*, March 1970, pp. 395–409.
57. *Steel Sheet Piling Design Manual*, U.S. Steel Corporation, Pittsburgh, Penn.
58. E. J. Yoder, *Principles of Pavement Design*, Wiley, New York, 1967.
59. Byron J. Prugh, "Densification of Soils by Explosive Vibrations," *ASCE Construction Journal*, March 1963.

60. D. D. Yoakum, "Winter Construction of Earthwork and Foundations," *Civil Engineering*, August 1967, p. 50.
61. *Standard Test Method for Density of Soil In-Place by the Sand-Cone Method*, ASTM D1556, American Society for Testing and Materials, Philadelphia, Penn.
62. *Standard Test Method for Density and Unit Weight of Soil In-Place by the Rubber Balloon Method*, ASTM D2167, American Society for Testing and Materials, Philadelphia, Penn.
63. *Standard Test Methods for Density of Soil and Soil-Aggregate In-Place by Nuclear Methods (Shallow Depth)*, ASTM D2922, American Society for Testing and Materials, Philadelphia, Penn.
64. Manufacturers of nuclear soil density testing equipment include (*a*) Seaman-Nuclear Corp., Milwaukee, Minn.; (*b*) Hydrodensimeter, Middlesex, U.K.; (*c*) Soil Test, Inc., Evanston, Ill.
65. *Standard Test Method for Density of Soil In-Place by the Drive-Cylinder Method*, ASTM D2937, American Society for Testing and Materials, Philadelphia, Penn.
66. "Landslides and Engineering Practice," *Special Report 29, Publication 544*, Highway Research Board, National Academy of Science, Washington, D.C., 1958.
67. Richard L. Handy and Wayne W. Williams, "Chemical Stabilization of an Active Landslide," *Civil Engineering*, August 1967, p. 62.
68. R. M. Koerner and J. P. Welsh, *Construction and Geotechnical Engineering Using Synthetic Fabrics*, Wiley, New York, 1980.
69. William B. Kays, *Construction of Linings for Reservoirs, Tanks, and Pollution Control Facilities*, 2nd ed., Wiley, New York, 1987.
70. Bruce A. Lamberton, "Revetment Construction by Fabriform Process," *ASCE Construction Journal*, July 1969.
71. *Foundation Facts*, No. 2, Raymond Concrete Pile Division, Houston, Tex., 1967; D. J. D'Appolonia, "Effects of Foundation Construction on Nearby Structures," Fourth Pan American Conference, San Juan, Puerto Rico, June 1971, American Society of Civil Engineers, New York.
72. *Soil-Cement Construction Handbook*, Portland Cement Association, New York.
73. "Who Pays for the Unexpected in Construction?" *Construction Journal*, ASCE, Committee on Contract Administration, September 1963.
74. William W. Moore, "Who Pays for Unforeseen Subsoil Conditions?" *Civil Engineering*.
75. F. A. Prentis and E. E. White, "Underpinning," Appendix B, *Legal Aspects of Underpinning and Foundation Work*, out of print.
76. G. F. Sowers, "Changed Soil and Rock Conditions in Construction," *ASCE Construction Journal*, November 1971.

Index

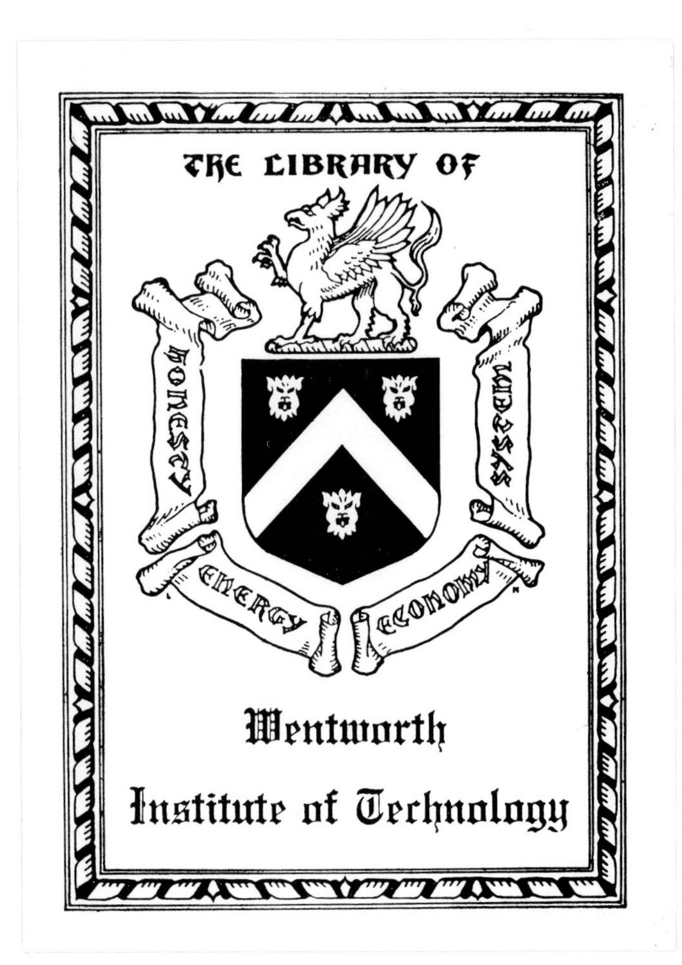
THE LIBRARY OF
HONESTY
SYSTEM
ENERGY
ECONOMY
Wentworth
Institute of Technology

Harold J. Rosen
CONSTRUCTION SPECIFICATIONS WRITING: PRINCIPLES AND PROCEDURES, Second Edition

Walter Podolny, Jr. and Jean M. Müller
CONSTRUCTION AND DESIGN OF PRESTRESSED CONCRETE SEGMENTAL BRIDGES

Ben C. Gerwick, Jr. and John C. Woolery
CONSTRUCTION AND ENGINEERING MARKETING FOR MAJOR PROJECT SERVICES

James E. Clyde
CONSTRUCTION INSPECTION: A FIELD GUIDE TO PRACTICE, Second Edition

Julian R. Panek and John Philip Cook
CONSTRUCTION SEALANTS AND ADHESIVES, Second Edition

Courtland A. Collier and Don A. Halperin
CONSTRUCTION FUNDING: WHERE THE MONEY COMES FROM, Second Edition

James B. Fullman
CONSTRUCTION SAFETY, SECURITY, AND LOSS PREVENTION

Harold J. Rosen
CONSTRUCTION MATERIALS FOR ARCHITECTURE

William B. Kays
CONSTRUCTION OF LININGS FOR RESEVOIRS, TANKS, AND POLLUTION CONTROL FACILITIES, Second Edition

Walter Podolny and John B. Scalzi
CONSTRUCTION OF CABLE-STAYED BRIDGES, Second Edition

Edward J. Monahan
CONSTRUCTION OF AND ON COMPACTED FILLS

Ben C. Gerwick, Jr.
CONSTRUCTION OF OFFSHORE STRUCTURES

David M. Greer and William S. Gardner
CONSTRUCTION OF DRILLED PIER FOUNDATIONS

James E. Clyde
CONSTRUCTION FOREMAN'S JOB GUIDE

Leo Diamant
CONSTRUCTION ESTIMATING FOR GENERAL CONTRACTORS

Richard G. Ahlvin and Vernon Allen Smoots
CONSTRUCTION GUIDE FOR SOILS AND FOUNDATIONS, Second Edition

B. Austin Barry
CONSTRUCTION MEASUREMENTS, Second Edition